Selected Data on Concentrated Acids and Bases[a]

Compound	Molarity	Density
Acetic acid	17.5	1.05
Ammonium Hydroxide	15.0	0.896
Hydrochloric acid	12.0	1.19
Nitric acid	15.0	1.43
Phosphoric acid	15.0	1.83
Sulfuric acid	18.0	1.84

[a] Values can vary with the age and source of the reagent.

Selected Atomic Weights

Element	Symbol	At. Wt.	Element	Symbol	At. Wt.
Aluminum	Al	26.982	Lithium	Li	6.939
Barium	Ba	137.34	Magnesium	Mg	24.312
Boron	B	10.811	Manganese	Mn	54.938
Bromine	Br	79.909	Mercury	Hg	200.59
Cadmium	Cd	112.40	Nitrogen	N	14.007
Calcium	Ca	40.08	Oxygen	O	15.999
Carbon	C	12.011	Phosphorus	P	30.974
Chlorine	Cl	35.453	Potassium	K	39.102
Chromium	Cr	51.996	Silicon	Si	28.086
Copper	Cu	63.54	Silver	Ag	107.870
Fluorine	F	18.998	Sodium	Na	22.991
Hydrogen	H	1.008	Sulfur	S	32.064
Iodine	I	126.904	Tin	Sn	118.69
Lead	Pb	207.19	Zinc	Zn	65.37

Crystallization Solvents (In approximately decreasing polarity)[a,b]

Solvent	Boiling Point	Freezing Point	Flash Point	Solvent	Boiling Point	Freezing Point	Flash Point
Water	100	0	—	Diethyl ether	35	−116	−45
Acetonitrile[c]	82	−44	1	Methylene chloride	40	−97	—
Methanol[c]	65	−98	11	Toluene	111	−95	4
Ethanol (100%)[c]	78	−117	13	Benzene	80	6	−11
Acetone[c]	56	−95	−20	Carbon tetrachloride	76	−23	—
Isopropyl alcohol[c]	82	−90	−18	Cyclohexane	81	7	−20
Acetic acid[c,e]	118	17	38	Petroleum ether[f]	(see footnote)		
Ethyl acetate	77	−84	−4	Hexane[g]	68	−95	−22
Chloroform	61	−64	—	Pentane[h]	36	−130	−40

[a]Order of decreasing dielectric constants.

[b]Temperatures are rounded to the nearest °C.

[c]Miscible with water.

[d]Both absolute (100%) and 95% ethanol are commonly used.

[e]Small amounts of water greatly increase the effectively polarity.

[f]Used in boiling ranges of 20–40°, 30–60°, 30–75°, and 60–110°; Fractions boiling above 60° are sometimes referred to as Ligroin.

[g]Known commercially as Skelly B.

[h]Known commercially as Skelly A; Skelly F, bp 30–60° can often be substituted.

Commonly Used Solvent Pairs for Crystallization

More Polar	+	Less Polar	More Polar	+	Less Polar
Acetone		Diethyl ether	Ethanol		Acetone
Benzene		Cyclohexane	Ethanol		Petroleum ether
Benzene		Petroleum ether	Ethyl acetate		Cyclohexane
Chloroform		Petroleum ether	Methanol		Methylene chloride
Chloroform		Carbon tetrachloride	Water		Ethanol
Diethyl ether		Hexane	Water		Acetone

Chemistry 203 | 204

Standard Scale Experiments

For Brookdale Community College

John A. Landgrebe

CENGAGE
Learning·

Australia • Brazil • Japan • Korea • Mexico • Singapore • Spain • United Kingdom • United States

Chemistry 203 | 204: Standard Scale Experiments, For Brookdale Community College

Senior Project Development Manager:
Linda deStefano

Market Development Manager:
Heather Kramer

Senior Production/Manufacturing Manager:
Donna M. Brown

Production Editorial Manager:
Kim Fry

Sr. Rights Acquisition Account Manager:
Todd Osborne

Theory and Practice in the Organic Laboratory with Microscale and Standard Scale Experiments : Fifth Edition
John A. Landgrebe

© 2005 Cengage Learning. All rights reserved.

For product information and technology assistance, contact us at
Cengage Learning Customer & Sales Support, 1-800-354-9706

For permission to use material from this text or product,
submit all requests online at **cengage.com/permissions**
Further permissions questions can be emailed to
permissionrequest@cengage.com

This book contains select works from existing Cengage Learning resources and was produced by Cengage Learning Custom Solutions for collegiate use. As such, those adopting and/or contributing to this work are responsible for editorial content accuracy, continuity and completeness.

Compilation © 2013 Cengage Learning

ISBN-13: 978-1-285-89543-7

ISBN-10: 1-285-89543-6

Cengage Learning
5191 Natorp Boulevard
Mason, Ohio 45040
USA

Cengage Learning is a leading provider of customized learning solutions with office locations around the globe, including Singapore, the United Kingdom, Australia, Mexico, Brazil, and Japan. Locate your local office at:
international.cengage.com/region.

Cengage Learning products are represented in Canada by Nelson Education, Ltd.

For your lifelong learning solutions, visit **www.cengage.com/custom.**
Visit our corporate website at **www.cengage.com.**

Printed in the United States of America

Preface

This text presents background material on both classical and modern laboratory techniques (microscale and standard scale) and instrumentation used in organic chemistry, as well as experiments that illustrate many of these techniques in addition to numerous important organic reactions and concepts. The book is intended for use in the laboratory portion of a one-year organic chemistry course for students entering the fields of biology, chemistry, chemical engineering, medicine, medicinal chemistry, or pharmacy. However, many of the experiments are quite suitable for a one-semester lab course. For pedagogical reasons, to provide flexibility, and so that the manual can serve as a useful reference for students doing undergraduate research, the chapters dealing with theory and techniques (Part I) remain separated from the experimental procedures (Part II). The overall emphasis is on learning fundamental techniques and procedures for handling, synthesis, separation, purification, and identification of organic compounds. Although this text contains more material than can be covered in a one-year laboratory course, this fact affords the instructor flexibility and provides resource material for advanced courses such as the undergraduate research experience. There continues to be a strong emphasis on safety and proper methods for handling and disposal of hazardous materials.

Although this fifth edition retains the general philosophy and format of previous editions, it is characterized by the extensive revision of many chapters in Part I, including the clarification and simplification of descriptions of concepts new descriptions of modern instrumentation, reorganization of topics in several chapters, and the inclusion of in-chapter exercises within all of the background chapters of Part I at locations where they represent a valuable learning tool for the student. Chapter-by-chapter details follow.

- Chapter 1, "Emergency Procedures and Safe Laboratory Practice," continues to reflect a wide range of information on first aid, fire safety, hazardous materials, waste management, and laboratory safety. Tables and references have been updated.
- Chapter 2, "General Laboratory Procedures," emphasizes the use of a sand bath in an electrically heated ceramic well, but also provides an extensive discussion of other commonly used heating methods, in addition to background in weighing, transferring, stirring, mixing, solvent evaporation, and the use of tubing and stoppers.
- Chapter 3, "Examination of Pure Compounds," has only minor changes in several descriptions of commonly employed techniques for measuring such things as melting points, boiling points, density, optical rotation, etc.
- Chapter 4, "Chromatography," includes an introduction to the polarity of molecules and has been clarified and updated to reflect changes in this field. The description of thin-layer chromatography has been rewritten to reflect modern practices with commercially available plates.

- Chapter 5, "Crystallization and Filtration," has a revised discussion on molecular structural effects on solubility, and other parts of the chapter have been reorganized, but continue to provide detailed procedures for crystallization on a conventional scale and microscale, the latter either with the conventional Craig tube or using the Pasteur pipet method developed by the author. Vacuum filtration emphasizes the use of a trapped diaphragm pump in preference to an aspirator.
- Chapter 6, "Extraction and Drying," continues to combine the techniques for extraction and for drying liquids; microscale extractions are done in small, inexpensive vials with a Teflon™ lined cap. New to this chapter are references to the use of stainless steel columns of activated alumina for the purification of solvents.
- Chapter 7, "Distillation and Related Techniques," has been extensively reorganized to include comparisons of conventional and microscale procedures. Several revised sections include the introduction and description of Raoult's law. Also discussed in this chapter are sublimation and steam distillation.
- Chapter 8, "Introduction to Sepectroscopy; Organic Structure Determination," provides basic information on the electromagnetic spectrum and the index of hydrogen deficiency, and gives an introduction to chemical methods for structure determination, but no longer contains a description of elemental analysis. From this short chapter it is possible to use any of the remaining chapters on spectroscopy in any order.
- Chapter 9, "Nuclear Magnetic Resonance Spectroscopy," is one of the most extensively revised chapters, with numerous 300-MHz spectra replacing the 60-MHz spectra of previous editions. Descriptions of important concepts have been rewritten. A brief introduction to 2D NMR spectra is also included.
- Chapter 10, "Infrared Spectroscopy," has been revised to include better description of the FTIR instrument, and explains how to take very good IR spectra on Teflon™ tape, thus avoiding the more expensive and high-maintenance salt plates or solution cells. Charts of characteristic infrared frequencies have been revised and greatly expanded.
- Chapter 11, "Mass Spectrometry," has revised information in the sections on chemical ionizations and desorption ionization methods, and an updated bibliography.
- Chapter 12, "Ultraviolet-Visible Spectroscopy," has not been changed.
- Chapter 13, "Special Laboratory Procedures," contains such topics as drying solids and gases, addition of liquids and solids to reaction vessels, inert atmospheres and anhydrous conditions, use of compressed gas cylinders, constant temperature control, solvent purification, hydrogenation, pyrolysis, and photochemistry. This chapter is a valuable resource to students who will do undergraduate research or are preparing for graduate work in chemistry. The only significant change is the addition of a description of the method of purifying solvents by pressuring them through activated alumina in stainless steel tubes.
- Chapter 14, "The Organic Chemical Literature," has been reorganized and updated to include current methods for using on-line databases such as SciFinder Scholar and STN on the Web.

New to Part II is the introduction of cooperative discovery labs in which students work together in teams, with the results of each team being shared by the entire lab section toward the end of that experiment. Another change is the emphasis on the use of an electrically heated sand bath as the principal heat source rather than the shielded heat lamp of previous editions. Instructions for microscale recrystallizations are now suitable for using either a Craig tube or the micropipet method developed by the author. Numerous additional, specific illustrations of equipment and apparatus have been included within the experiments so that students do not have to continually refer to more general figures in various parts of the background chapters earlier in the text. Microscale procedures are compared with their conventional scale counterparts within the appropriate chapter for a particular technique, and are not tied to a speific type of glassware or kit. In fact, many microscale operations can be performed with conventional semi-micro 𝔉 14/20 joint) apparatus. To help guarantee a successful microscale experience, experiments are designed so that liquid products are normally obtained in quantities of at least 200 μL and solid products in quantities of no less than 50 mg and often more. Although microscale experiments have the advantages of shorter procedure times, fewer hazards, decreased supply costs, and less hazardous waste generation, they fail to expose students to some important techniques and apparatus, such as the use of separatory funnels, conventional distillation procedures, and standard scale recrystallizations, that remain an important part of the everyday world of organic chemistry. It is for the latter reason that some standard scale experiments are included in this manual, although even these experiments never exceed a few grams of starting material.

Part II contains a variety of experiments to introduce and illustrate important and basic laboratory skills and techniques, some important reactions commonly encountered in the corresponding lecture course, and many fundamental mechanistic concepts. In addition to new microscale experiments, many of the experiments from the previous edition have been improved, a few have been deleted, and each experiment has been carefully tested. In addition, there is an adequate variety of experiments to provide flexibility and choices for the one-year organic laboratory course. The experiments involving qualitative analysis include only a modest number of chemical tests and observations, but with a much heavier emphasis on spectroscopy than in the previous editions.

As in past editions, answers to selected exercises in Part I can be found in the Appendix. Students are expected to complete a formal write-up of each experiment as described in "Introduction to the Organic Laboratory," or as recommended by their instructor.

An Instructor's Manual is provided with answers to all exercises within the experiments and other information helpful for organizing the laboratory course and preparing for each experiment.

I should like to thank my colleagues for their encouragement and help, my teaching assistants who made many valuable suggestions, and my students for their interest and enthusiasm. Special thanks go to Professors David Benson, Albert Burgstahler, and Paul Hanson for numerous insightful discussions and improvements to the experiments, to Bruce Johnston, our Laboratory Educational Technician, for the organic labs and for his help with "A Word to the Instructor" as well as suggestions about equipment that have improved the experiments, and to Research Assistant Sarah Neuenswander for her help obtaining several new NMR spectra, including those for the 2D NMR section of Chapter 9. I especially appreciate working with the thoroughly professional staff at Brooks–Cole Publishing Company, and I am indebted to the expert editorial and production services of Matrix Productions. I wish to express my gratitude to Aldrich Chemical Company, Milwaukee, Wisconsin, and Perkin-Elmer Corporation, Norwalk, Connecticut, for permission to reproduce the nuclear magnetic resonance and infrared spectra found in Chapters 9 and 10, and to Brucker, San Jose, California, for permission to use their ^{13}C chemical shift chart in Chapter 9. Finally, to my wife, Carolyn, my grateful thanks for her forbearance and love during the preparation of the manuscript.

A Word
to the Instructor

The experiments in this manual represent a mix of microscale (<500 mg) and standard scale (but <2 g) experiments, and provide students with a range of important laboratory experiences. The microscale experiments have been designed so that they can be done with conventional semimicro (e.g., $\mathbb{S}$ 14/20) apparatus or with commercial kits such as those distributed by Ace, Chemglass, Corning, Kontes, Reliance, or Wheaton. Depending on the exact glassware items that are available at a given college or university, it may be necessary for the instructor to sketch a slightly modified version of what is pictured in this text and hand that out to students or post it in the laboratory room.

The heat source recommended in this text is an electrically heated sand bath, but a shielded 250W infrared lamp connected to a Variac (or less expensive solid state or proportional control) will also work, as described in Chapter 2. When using a magnetic stirrer with a sand bath, test the magnetic stir bar in the flasks that will be used, to be certain that they stir smoothly. Recrystallization is done with a Craig tube or the convenient Pasteur pipet method. (However, if the Pasteur pipet method is employed, a heat lamp is required.)

Following is a list of equipment recommended for doing the microscale and standard scale experiments described in this text. If the expense of supplying certain equipment for individual student drawers becomes too great, this equipment can be treated as community property, and either checked out on a daily basis as needed or marked with a laboratory station number (etched or baked enamel) and placed into a wall-mounted rack.

Individual Student Equipment

More expensive items or items used less frequently are designated for check-out from a stockroom to reduce costs.

1. Aluminum plate for IR measurements on Teflon[TM] tape ($1\frac{7}{8}'' \times 3'' \times \frac{1}{16}''$ thick with a $\frac{5}{8}''$ hole at center; see Ch. 10; 1/student)
2. Beakers (400 mL, 250 mL, 150 mL or 100 mL, 1 ea/student)
3. Ceramic well for sand bath (100 mL) or heat lamp, 250W, red-glass infrared, with shield[1] (1/student)

---

1. Swivelier shield #22093, Nanuet, NY, 10954. Some type of heat lamp must be used if the Pasteur pipet method for microscale recrystallizaion is selected.

4. Chromatography column (15 × 250 mm with Teflon™ stopcock; **check out**)
5. Condenser (West or comparable type, ℥ 14/20 female joint; 1/student)[microscale] **(check out)**
    Condenser (24/40, for reflux or distillation; 1/student) **(check out)**
6. Craig tube (if that method is selected for recrystallizations)[microscale] **(check out)**
7. Cylinders, graduated (25 mL, 100 mL; 1 ea/student)
8. Cylinder, graduated (10 mL or 5 mL; 1/student)[microscale]
9. Distillation adapter, ℥ 24/40 (for male end of condenser; 1/student; optional; **check out**)
10. Distillation column, fractional (made from a condenser with copper sponge; 1/student; **check out**)
11. Distillation head (simple, for fractional distillation; **check out**)
12. Distillation unit, Hickman (commercial unit or homemade, as in Figure 7.6;[2] 1/student)[microscale]; cloth strip and wire (2 cm × 10 cm, to wrap around condensing surface of Hickman still; 1/student)
13. Distillation unit, semimicro (**check out**)[3]
14. Flask, filter (250 mL; **check out**)
15. Flask, filter (125 mL; 1/student)[4, microscale]
16. Flasks, pear-shaped (℥ 14/20; 2/student; 10 mL and 25 mL)[microscale]
17. Flasks, Erlenmeyer (250 mL, 1/student; 125 mL, 1/student; 50 mL, 2/student; 25 mL, 4/student)
18. Flask, round-bottomed, single-necked, ℥ 24/40 (100 mL, 250 mL, 500 mL; 1 ea/student)
19. Funnel, Büchner (5.5 cm; 1/student)
20. Funnel, Hirsch (42 mm top dia, plate dia 11.5 mm, with stopper or adapter for filter flask or side-armed test tube; **check out**)[microscale]
21. Funnel, Hirsch (55 mm top dia, plate dia 16 mm, with stopper or adapter for filter flask; 1/student)[microscale]
22. Funnel, powder (75 mm top dia; 1/student)
23. Funnel, short stem (50 or 75 mm top dia; 1/student)
24. Funnel, separatory, Squibb (60 mL or 125 mL; **check out**)
25. Funnel, separatory, Squibb (250 mL or 500 mL; **check out**)
26. Medicine dropper (with bulb, serves as microscale drying tube; 1/student)[microscale]
27. Pipets, graduated (1 mL and 5 mL; **check out**)
28. Spatula, microscale[5] (1/student)[microscale]
29. Spatula, semimicro (stainless steel; 1/student)
30. Stirrer, magnetic (for 10-mL pear-shaped flask; **check out**)[microscale]
31. Stirring bar (Teflon™ covered, maximum length = 22–23 mm, but does not have to be the special higher priced triangle-shaped type; **check out**)
32. Sublimator, *ca.* 32 mm o.d. × 140 mm tall; see Figure 7.21b (**check out**)
33. Syringe, plastic, 1 mL (with tight-fitting plastic plunger, serves as dropping funnel; **check out**)[6, microscale]
34. Test tubes (several of assorted sizes, including 6 × 50 mm)
35. Test tube, side-armed (20 × 150 mm, serves as vacuum oven; **check out**)

---

2. This still can be prepared from a 10-mL blank pear-shaped flask from Chemglass, Vineland, NJ (1-800-843-1794 when outside New Jersey or 1-609-696-0014) and Pyrex™ tubing with o.d. 10 mm, ca. 2.5 cm at the top and 3 cm at the bottom). The receiver can handle volumes from 200 μL to 1 mL.
3. The Kontes short-path still or another comparable apparatus can be used.
4. A side-armed test tube can be substituted.
5. For example, Fischer #21-401-25A or B
6. Air-Tite Products, Vineland, NJ 1-800-257-5318, CA. $0.85/ea.

36. Thermometer (0–150°C, nonmercury)
37. Thermometer (0–400°C, mercury; **check out**)
38. Tube, drying (150 mL; 1/student)
39. Tube, glass (4–5 cm × 7 mm) with rubber stopper to fit ℥ 14/20 joint (for solvent evaporation; 1/student) microscale
40. Tweezers (1/student)microscale
41. Vials (for extractions and recrystallizations; 1 dram, 2/student; 0.5 dram, 2/student) with screw caps (Teflon[7] or polyethylene lined)microscale
42. Watch glass (1/student)microscale

General Laboratory Equipment

1. Balances, top loader (1-mg readability, with shield around pan; 1/10 students)
2. Boiling chips (e.g., boileezers) and sticks
3. Burets (50 mL) for dispensing certain reagents in hood
4. Clamps, rings (2″, 3″, 4″, 5″), pinch clamps (latter could be in each drawer)
5. Copper wire (20 gauge; 1 spool/lab)
6. Cotton (washed with hexane and methanol; several small bottles/lab)
7. Filter aid
8. Filter paper (to fit Hirsch, Büchner, and other funnels)
9. 50-μL gas chromatography syringe (to load capillary tube for bp determination; several/lab)
10. Glass wool (Pyrex™)
11. Labels
12. Melting-point capillaries
13. Microburners (1 or 2 in hood)
14. Pasteur pipets (large box/lab, both the 14- and 22-cm lengths)
15. pH paper
16. Pipets, automatic delivery (20 μL–200 μL; 1/10 students)
17. Pipets, automatic delivery (200 μL–1000 μL; 1/10 students)
18. Pliers with wire cutter (2–4 pairs/lab)
19. Ruler, plastic, 1 ft, metric and English (several/lab)
20. Silicone oil (in dropper bottles)
21. Sponges
22. Spot plates (black and white porcelain; at least several/lab)
23. Stopcock grease
24. Teflon™ tape (1/2″ × 600″ from Ace Hardware store) for IR measurements (See Ch. 10)[8]
25. Thin-layer chromatographic equipment (wide-mouthed jars, 80 mm high, with screw cap; 3/student; commercial TLC plates, *ca.* 2.5 × 7.5 cm; 6/student)
26. Variable transformer (Variac) or less expensive solid-state or proportional control[9] (1/student station)
27. Weighing paper (glazed; 1 box/balance)
28. Waste containers for nonchlorinated solvents, chlorinated solvents, other waste materials

---

7. Teflon™-lined cap for 0.5-dram vial, Fischer #02-883-3AA, 8-mm size; for 1-dram vials, #02-883-3A, 13-mm size.
8. Teflon™ tape from Fisher is thicker, more expensive, and does not work well.
9. Laboratory Craftsmen, Inc., Beloit, WI (1-608-362-2255).

Chemicals and Reagents for General Use[10]

1. Acetone (for drying and cleaning, in squeeze bottles)[11]
2. Calcium chloride (granular, anhydrous)
3. Decolorizing charcoal (Norite pellets, 0.8×3 mm, Aldrich)
4. Detergent solutions (Alkanox or Micro)
5. Hydrochloric acid (concentrated, corrosive cabinet until needed)
6. Hydrochloric acid ($3M$, corrosive cabinet until needed)
7. Magnesium sulfate (anhydrous)
8. Methylene chloride (solvent cabinet until needed)
9. Nitric acid (concentrated, corrosive cabinet until needed)
10. Salt (rock or granular, for ice-salt baths)
11. Sodium bicarbonate (saturated solution)
12. Sodium chloride (saturated solution)
13. Sodium hydroxide ($3M$)
14. Sodium sulfate (anhydrous)
15. Sulfuric acid (concentrated, corrosive cabinet until needed)

An Instructor's Manual can be requested from the publisher, which provides the following additional information: (1) chemicals or special supplies required for each experiment; (2) experiments subdivided by functional groups undergoing reaction or being produced; (3) a listing of the experimental techniques used in each experiment; (4) reaction sequences within and among various experiments; (5) answers to the exercises within each experiment; and (6) a listing of discovery and cooperative-discovery experiments.

---

10. Special solvents and reagents to be supplied in appropriate quantities as needed.
11. According to OSHA (Ocupational Safety and Health Act) regulations, the maximum amount of this solvent that can be stored in glass or approved plastic containers is 1 qt and in metal or safety cans is 1 gal.

CHAPTER 2

General Laboratory Procedures

2.1 Weighing and Transferring on a Microscale

Success on a microscale depends on the use of techniques and apparatus that minimize handling. Every time a solid or liquid is transferred from one container to another, there are opportunities for loss; therefore, any such *transfers must be kept to a minimum* and performed carefully. A neat and well-organized bench top is essential when dealing with small amounts of materials. Make certain that all containers are carefully labeled. For purposes of this discussion, microscale is considered <500 mg of solid or 500 μL of liquid.

Reactions can be carried out in a small (5- or 10-mL) pear-shaped flask (Figure 2.1a), or in a heavy-walled, conical reaction vial (Figure 2.1b).[1] When a pear-shaped flask is not clamped to a support rod, it should be supported in a small cork ring or placed into a small beaker.

1. These vials come in several sizes, have a short standard-taper female joint, and a threaded cap with an opening and an O-ring to make a firm connection with other apparatus (e.g., a condenser). See Section 2.6 and Figure 2.24 for more details.

Solids

Use a sensitive top-loading balance to weigh solids to the nearest milligram. When a solid is to be weighed into an empty flask, weigh the empty flask first (supported on a cork ring or in a small beaker).[2] Then add a small amount of the solid with a spatula and check the new weight. To save time at the balance, you may stop when your weight is within a few milligrams of that specified. Adjustments in the amounts of other reagents won't be necessary if they are present in excess.

2. The flask is said to be tared.

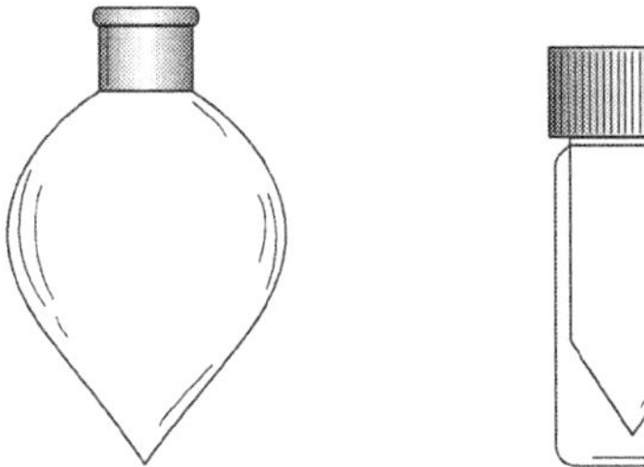

Figure 2.1 Microscale reaction vessels

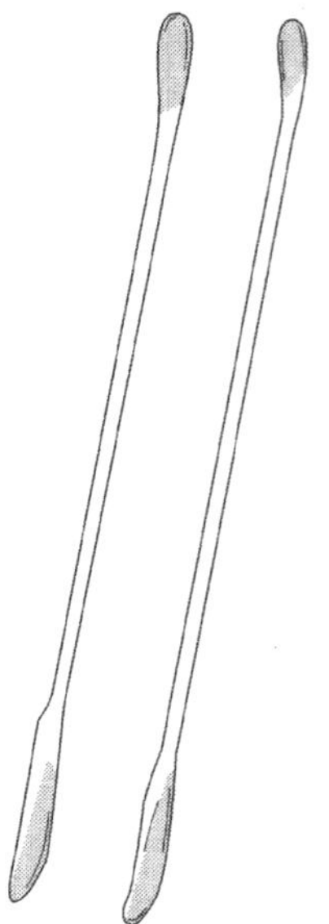

Figure 2.2 Stainless steel microspatulas

Although there are times when solids can be weighed onto glazed paper followed by transfer into a second container, great care must be taken to avoid significant losses. A stainless steel microspatula (Figure 2.2) can be used to help minimize losses during the transfer of a solid.

Liquids

Automatic delivery pipets, Figure 2.3, should be used for accurate (±2%) delivery of fixed amounts of liquid, often variable over a specified range of volume, and many of these pipets can be read to at least 1 μL (10^{-3} mL). They draw the liquid into a polyethylene tip that can eventually be discarded. The following procedure should lead to reproducible results for liquids that are not very viscous.

- Depress the button at the top of the pipet until resistance is encountered (first stop).
- Dip the tip a few mm below the surface of the liquid and carefully release the button to draw up the liquid. *Never let the button release with a snap.* There should be no air bubbles in the tip; if there are, try again or switch to a new tip. If there is any liquid on the outside of the tip, wipe it carefully with an absorbent tissue.
- Then put the tip of the pipet into the container into which delivery is desired and push the button slowly past the first stop. Touch the tip to the inside of the container.

Automatic pipets should always be stored in a vertical position so that no liquid can come into contact with the seals.

When accurate delivery is not needed, small amounts of liquids can be conveniently transferred with a **Pasteur pipet**, Figure 2.5a, or (if the liquid is not corrosive) a 10-μL gas chromatography syringe, Figure 2.4. *For a typical 7-mm (OD) Pasteur pipet, 4 mm of length corresponds to approximately 0.1 mL, a useful number to remember.* For liquids with a high vapor pressure, which tends to make them squirt from the Pasteur pipet, the **filter pipet**, Figure 2.5b, gives better control of delivery. The filter

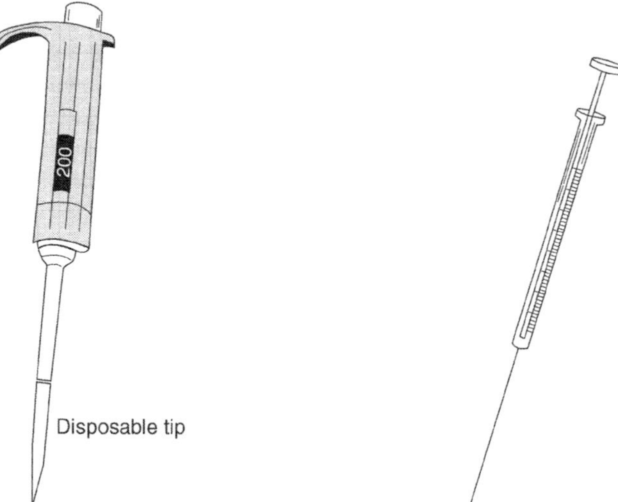

Figure 2.3 Automatic delivery pipet   **Figure 2.4** Gas chromatography syringe

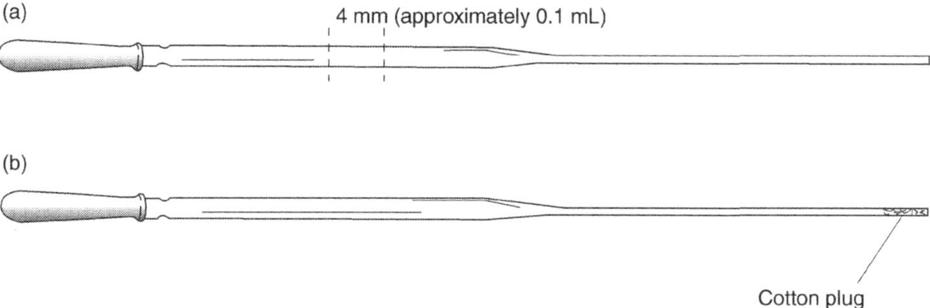

Figure 2.5   (a) Pasteur pipet; (b) filter pipet

pipet contains a very small tuft of cotton that has been pushed through the pipet to the tip with the aid of a piece of straight 20-gauge copper wire. (Straighten the wire by holding it firmly at both ends with pliers, snapping it, and cutting off the ends.) Some practice in selecting the correct amount of cotton is usually needed to become proficient.[3] Filter pipets have a variety of uses, described in Chapter 5 (Crystallization and Filtration) and Chapter 6 (Extraction and Drying).

  Liquid quantities of several mL can be dispensed from a 5-mL or 10-mL graduated cylinder or from a buret. The latter technique, when done in a hood, is especially useful for corrosive liquids or liquids with obnoxious odors.

3. For the best results use cotton that has been rinsed with small amounts of methanol and hexane, to remove undesired impurities, and then dried.

EXERCISES

1. What is the simplest device for dispensing approximately 0.3 mL of an organic solvent into a microscale reaction vial?

2. Under what circumstances would it be useful to use a filter pipet rather than a Pasteur pipet for transferring a small amount of liquid?

3. If an automatic pipet has an accuracy of $\pm 2\%$, what would the range of possible volumes be (to the nearest 0.1 μL) for a 50-μL delivery?

4. The method described for using an automatic delivery pipet works particularly well for liquids that are not too viscous. Too much viscous liquid remains in the tip and leads to too little liquid being delivered. How would you redesign this procedure so that an exact amount of a viscous liquid could be dispensed?

2.2 Transfer of Solids and Liquids on a Macroscale

Solids

As with microscale handling procedures, gram quantities of solids can be weighed into a **tared** (preweighed) reaction vessel or can be weighed onto nonabsorbent, glossy paper on a top-loading balance and then transferred to a glass container. A large amount of solid material should be poured into a flask (or any other narrow-necked

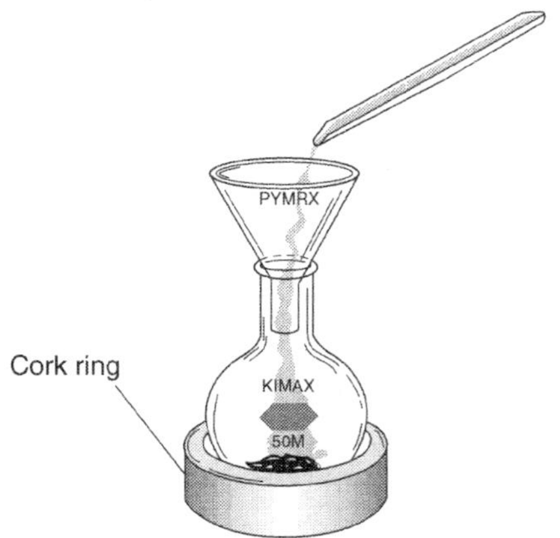

Figure 2.6   Use of a powder funnel

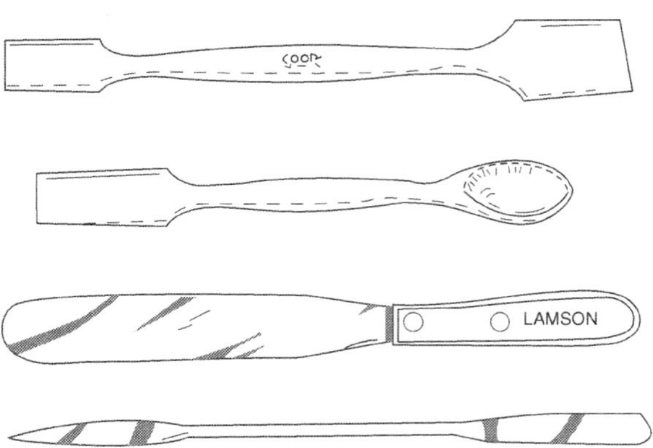

Figure 2.7   Stainless steel and porcelain spatulas

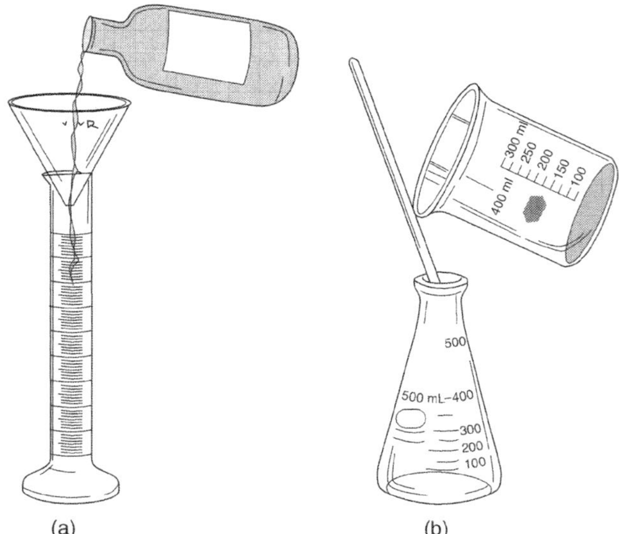

Figure 2.8   Use of funnels and stirring rods when pouring liquids

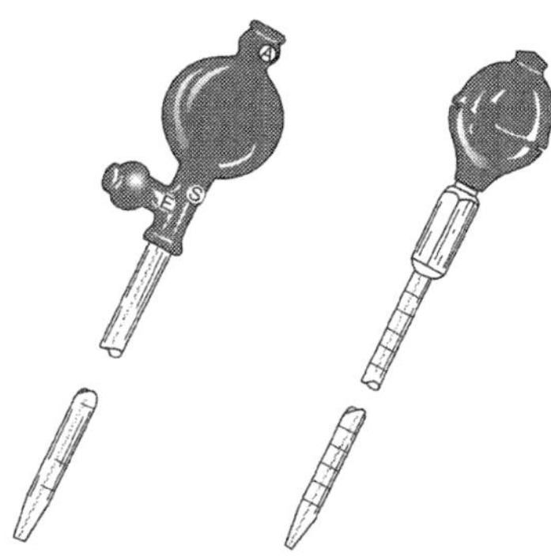

Figure 2.9   Safety bulbs for conventional pipets

container) with the aid of a powder funnel as shown in Figure 2.6. A stainless steel or porcelain spatula (Figure 2.7) can be used to minimize losses during transfer.

Liquids

When liquids are poured into a container with a small opening, the use of a funnel (Figure 2.8a) or a stirring rod (Figure 2.8b) will help eliminate spills and losses. Liquids should always be poured in a hood where potentially harmful vapors will be properly exhausted. If conventional pipets are used to make a quantitative transfer, a rubber bulb (*never the mouth*) must be employed (Figure 2.9).

5. What is the physical distinction between an ordinary funnel and a powder funnel?

6. Explain the procedure and proper precautions for drawing liquid into a conventional pipet.

7. Describe how a stirring rod can be used to help transfer a liquid from a beaker to an Erlenmeyer flask.

2.3 Stirring and Mixing

Many reaction mixtures that are heterogeneous, or require high dilution or very even heating, must be efficiently stirred. Some devices for automatic stirring are shown in Figure 2.10. For nearly all small-scale reactions and for reaction mixtures that are not too viscous or do not contain extensive amounts of heavy precipitates, **magnetic stirring** (Figure 2.10a) provides the ultimate in simplicity. A Teflon™-coated stirring bar, available in a variety of sizes and shapes, is placed within the vessel, which is then seated or clamped on top of the magnetic stirring motor. The speed of rotation can be varied and some magnetic stirrers come equipped with a top surface that is also a hot plate (see next section for a discussion of heating devices).

Two commonly used devices for overhead stirring are the Hershberg and sleeve types shown in Figure 2.10b and c, respectively. The bearing of the inexpensive **Hershberg stirrer** consists of a rubber tube that seals against the glass shaft of the stirrer and is lubricated by a few drops of silicone oil placed in the cup at the top.[4] When properly adjusted these stirrers can run overnight without additional oil. The Hershberg stirring paddle consists of twisted 18-gauge tantalum wire. The less expensive and more

4. If oil leakage from the cup is a problem, the seal can be made tighter by use of a rubber band on the outside of the tubing.

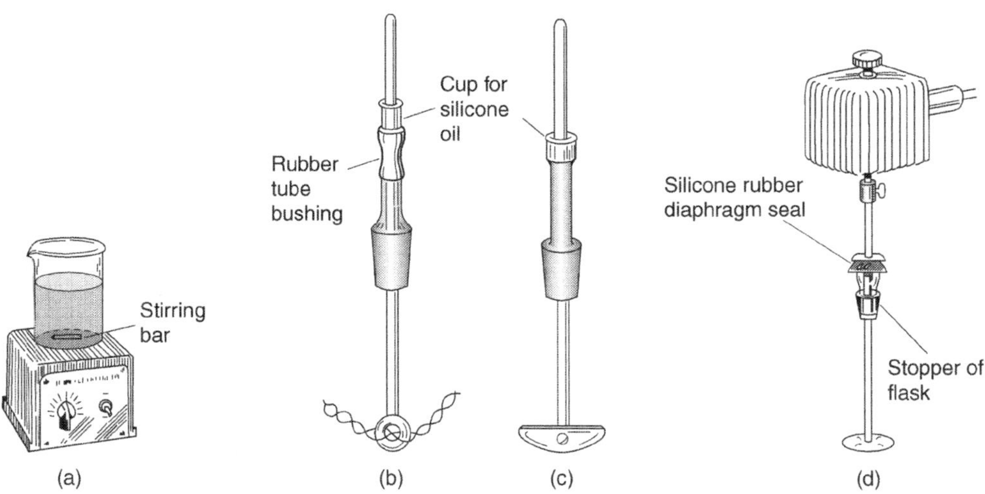

Figure 2.10 Automatic stirring devices: (a) magnetic; (b) Hershberg; (c) sleeve; (d) vibrating (Vibromischer)

available nichrome wire is sometimes used in place of tantalum, but is not as chemically inert.[5]

The **sleeve stirrer** has a ground-glass bearing lubricated with a little silicone oil or a Teflon™ bearing that doesn't require lubrication. These stirrers are often used in conjunction with a glass or Teflon™ paddle. *Teflon™ should never be used in a reaction mixture containing an active metal such as sodium or potassium.* This type of stirrer is expensive and must be carefully aligned to avoid undue wear on the shaft and bearing. Because some vibration during stirring is impossible to avoid, it is important to clamp both the bearing and the neck of the flask (so that the pressure of the clamp will keep the joint together) or to use the simple device shown in Figure 2.11 to keep the joint between the stirrer and the flask tightly connected. Even a small leak can result in significant loss of the solvent being employed in the reaction.

Reactions that require vigorous mixing can be done conveniently with a vibrating metal (stainless steel) or glass plate on the bottom of a shaft that passes through a simple diaphragm seal (Figure 2.10d). The plate contains perforations that, because they are tapered, force liquid through with great turbulence as the plate vibrates. The degree of agitation of the solution can be controlled by an adjustment on the motor.[6]

EXERCISES

8. What is the metal preferred for use as the wire loop of a Hershberg stirrer?

9. What is the most convenient method for stirring a reaction mixture in a small flask?

10. Teflon™ should not be allowed to come into contact with _________ or _________.

11. Briefly describe the purpose of a vibrating mixer or stirrer.

2.4 Heating and Temperature Control

Heating Devices

Several types of laboratory heating devices are shown in Figure 2.12. For all of the electrical devices, a **variable transformer (Variac**, Figure 2.13) provides an easily controlled source of voltage to vary the temperature.[7] With the exception of the heat lamp, none of these electrical devices should ever be plugged directly into a 110V outlet!

A **labjack** (Figure 2.14) under heating devices (a) or (c)–(e) will allow them to be safely raised or lowered under a flask that has been clamped to a support rod.

A ceramic-well heater (Figure 2.12a) about two-thirds full of sand makes a convenient **sand bath** for heating small flasks and other microscale apparatus.[8] Use clean (washed) sand and clamp a thermometer with the bulb immersed below the surface. Because sand is a poor heat conductor (it takes a long time to change temperature and is hotter at the bottom of the sand than at the top), it is best to heat the sand somewhat above the temperature needed and then control the temperature in a glass vessel by immersing the vessel to a greater or lesser extent in the hot sand. Magnetic stirring will work through a sand bath of this type.

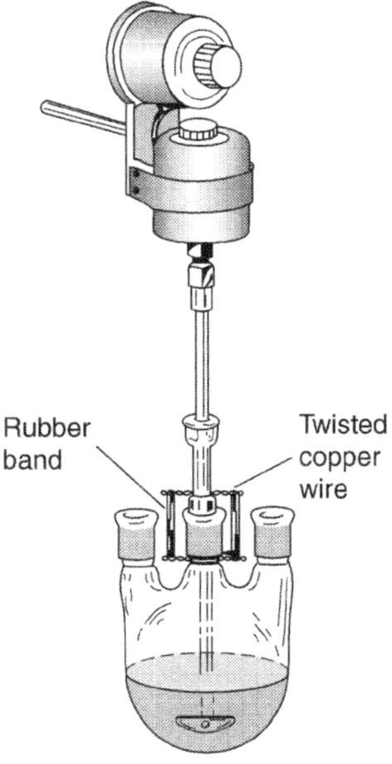

Figure 2.11   Simple device to stabilize stirrer attachment

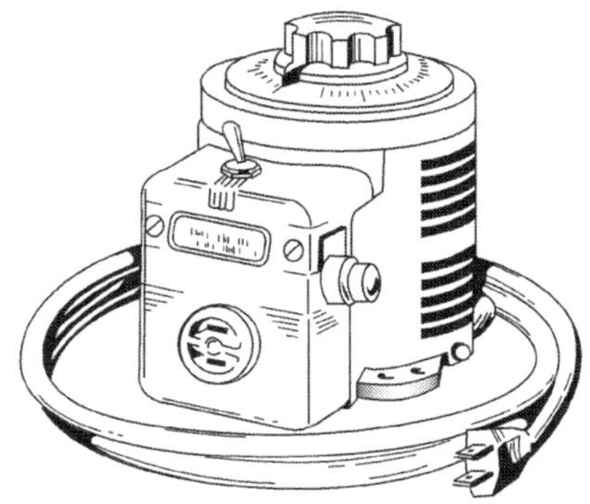

Figure 2.13   Variable transformer or Variac

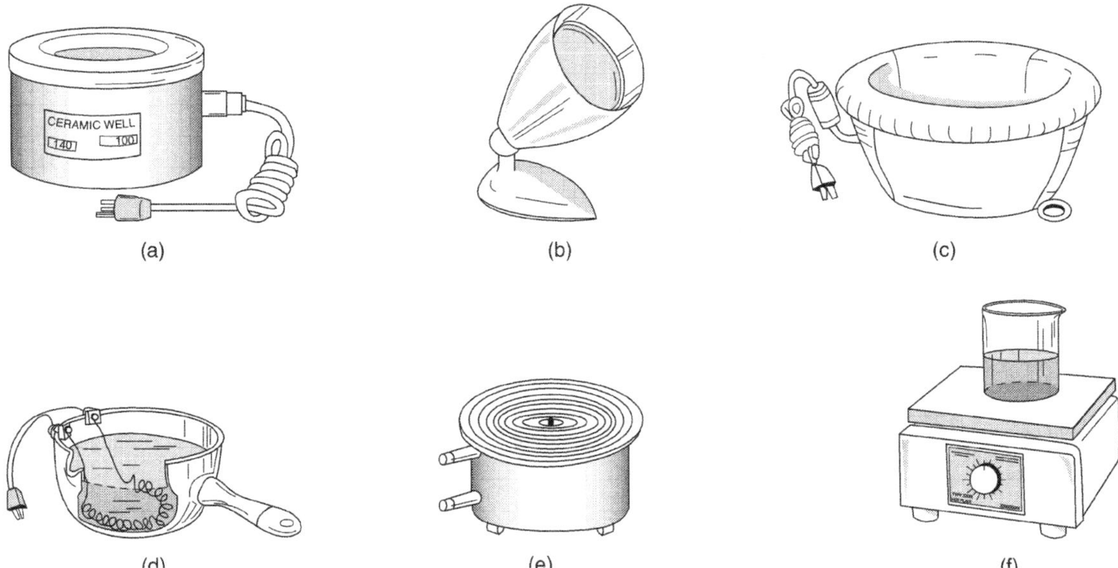

Figure 2.12   Heating devices: (a) sand bath in ceramic well; (b) shielded heat lamp; (c) heating mantle; (d) oil bath; (e) steam bath; (f) hot plate

A comparatively safe, convenient, and flexible heat source is a shielded **infrared heat lamp** (Figure 2.12b).[9] A method applicable over a wide range of temperatures, when stirring is not required, is to place a shielded heat lamp vertically upward with a set of metal rings (normally used on a steam bath) on the top as in Figure 2.15a.

9. The standard size lamp is 250 watt; the shield is Swivelier #22X-0376WH. One Amy Kay Parkway, Kingston, NY 12402.

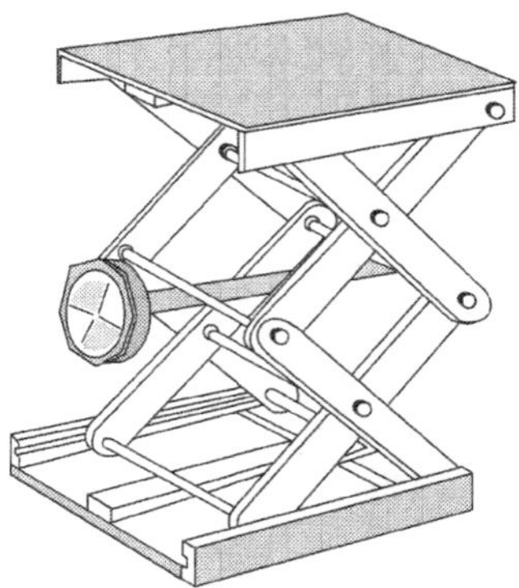

Figure 2.14 Labjack

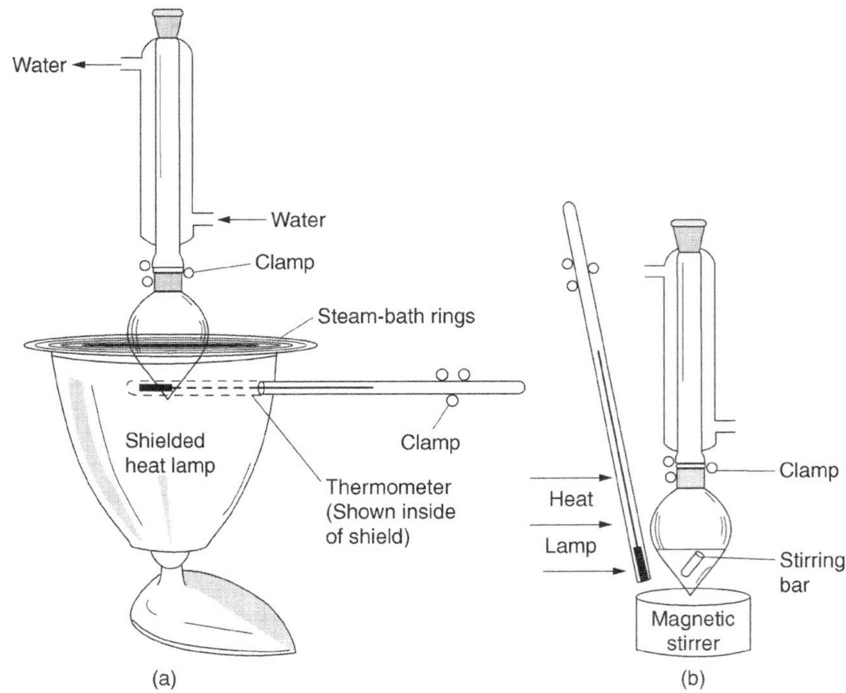

Figure 2.15 Arrangements with shielded heat lamps: (a) vertical air bath; (b) heating and stirring (aluminum foil used to direct and confine heat not shown)

A small hole drilled near the upper edge of the aluminum shield accommodates a thermometer, which is then clamped horizontally to hold it in place. The bulb of the thermometer should be alongside the pear-shaped flask, with the latter clamped so that it sits in the smallest opening of the ring set. If the smallest ring is missing, aluminum foil or glass wool can be used to seal around the flask. In this manner the dead air space around the flask serves as an **air bath**, and the temperature is readily controlled by a variable transformer. *Care should be taken not to spill flammable liquids onto the hot surface of the infrared bulb.* When the shielded heat lamp is used in the vertically upward orientation, the steam-bath rings (shown in Figure 2.15a) should *always* be present to protect against accidental breakage of the lamp if a cold liquid is spilled above it.

If stirring and heating are required, a small magnetic stirring bar is placed into the flask, which is then clamped above the magnetic stirrer. The heat lamp is directed horizontally, and aluminum foil is used to train the heat on the flask from behind and to shield heat from the condenser. A thermometer should be placed alongside the flask to monitor the temperature. The apparatus is illustrated in Figure 2.15b.

Heating mantles (Figure 2.12c) are made from woven fiberglass cloth and an embedded heating element controlled by a transformer. *Do not apply more than the maximum voltage indicated on the attached tag or label.* Because they are expensive and are designed for a specific size of flask (beginning with 50 mL), a number of different sizes are needed to heat a range of flask sizes.[10] Heating mantles can be conveniently supported under the flask by a glass ring or a labjack, and will normally

10. In fact, a specific mantle can usually be adapted for use with flasks one or two sizes smaller than the rated size by surrounding the flask with glass wool.

work with magnetic stirring devices. More expensive mantles are equipped with an embedded thermocouple that can be attached to a meter that displays the temperature. Note that magnetic stirrers will operate through conventional heating mantles.

The **oil bath** (Figure 2.12d) provides easily controlled and uniform heating for small to medium-sized flasks. (Oil baths for flasks above 500 mL are awkward and dangerous.) The bath consists of a porcelain crucible containing a nichrome heating coil mounted with screw clamps as shown in Figure 2.12d.[11] The cord from the variable transformer is attached directly to the screw clamps. More uniform heating can be achieved by placing a magnetic stirring bar in the oil and standing the crucible on a magnetic stirrer. Care should be taken to prevent the stirring bar from contacting the heating coil.

A few comments about the special hazards of oil baths are appropriate. Dow Corning 550 silicone oil (expensive) is recommended because it is not combustible and can be used for extended periods at temperatures as high as 230°C without serious decomposition. Other oils must be used with great caution because overheating could result in a fire. For example, inexpensive paraffin or mineral oil should not be used at temperatures above 175°C. *Serious burns can result from spattering of oil caused by water falling into the bath, sinking, and evaporating with explosive force.* Thus, special caution must be observed when this type of heating device is being used. Water hoses to condensers must be carefully checked and if water droplets are observed in the bottom of a cold oil bath, the oil should be drained and replaced before the bath is heated.

Although the **steam bath** (Figure 2.12e) heats rapidly and has some obvious safety advantages when flammable solvents are being heated, temperatures are restricted to 100°C or less and temperature control is minimal. The increase in humidity from the use of a large number of steam baths in a laboratory not only decreases comfort, but can make it difficult or impossible to work with moisture-sensitive compounds.[12]

Hot plates (Figure 2.12f) can be useful for warming aqueous solutions in flat-bottomed vessels, but they are not recommended for heating flammable solvents in open vessels. Most laboratory hot plates are equipped with thermostats that can spark and thus ignite flammable vapors. Hot plates with built-in magnetic stirrers are also available.

11. Approximately 10 ft of 26-gauge wire is sufficient.

12. Because of cost, maintenance problems, and humidity concerns, many modern instructional and research labs are no longer equipped with steam lines.

Liquids at Reflux

The simplest and most widely used procedure for maintaining a constant reaction temperature is to carry out the reaction in a suitable solvent at its boiling point (Figure 2.16). The solvent is said to be at **reflux**, which means that all of the solvent vapor is being condensed and returned to the boiler flask. Any of the previously mentioned heat sources can be used. Changes in the heat input to the flask or in the heat produced by a reaction will cause the solvent to boil more or less vigorously, but will not alter the temperature. The flask on the left in Figure 2.16 has a more efficient condenser (Friedrichs) that would be especially useful for low-boiling solvents. *To promote smooth boiling, boiling chips or sticks should be added to the liquid before it is heated, or the liquid should be stirred (magnetic stirrer, etc.) while it is being heated.*

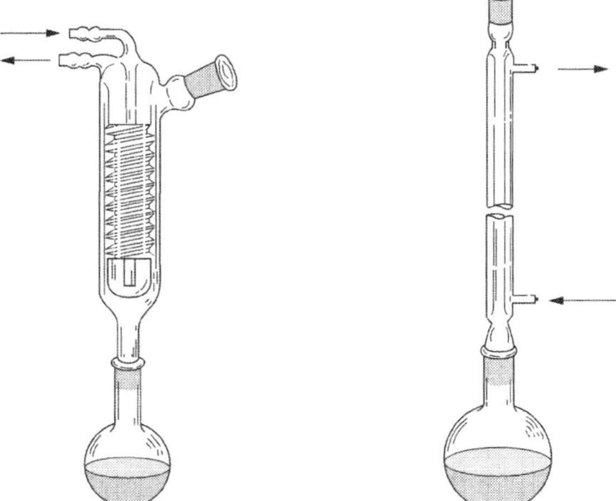

Figure 2.16 Maintaining a liquid at reflux (heat sources not shown; arrows show flow of cooling water)

12. Indicate safety considerations for the use of a hot plate in a laboratory.

13. Which of the heating devices listed can be used with a magnetic stirrer?

14. What are special hazards of using oil baths as heating sources?

15. Briefly describe what is meant by "maintaining a liquid at reflux."

16. Why is it dangerous to add a boiling chip to a hot organic liquid?

17. Briefly describe the purpose of each of the following items: (a) Variac; (b) labjack.

2.5 Solvent Evaporation

Reaction products are frequently obtained as solutions in an organic solvent. Although distillation (Chapter 7) is a conceivable means for removing the solvent, it has the disadvantage of being slow, and the residue, if subjected to an extended period of heating, may decompose. What is needed is a means of rapidly concentrating solutions under mild conditions. **Rotary evaporators**, such as those shown in Figure 2.17, are one answer to this problem.

To use a rotary evaporator, the flask (at the right in the figure) is half-filled with the reaction solution and rotated partially submerged in a warm water bath. A vacuum is kept on the system by means of a diaphragm pump[13] with a dry ice–cooled trap or an aspirator connected to a cold trap. The solution is continuously dispersed on the warm, inner surface of the rotating flask and evaporates rapidly under these relatively low-temperature conditions. Solvent vapors are condensed and collected in a receiver, that can be cooled. Some apparatus is equipped to allow the evaporator flask to be refilled without any interruption of the evaporation process (see stopcock B in Figure 2.17).

A simplified method of removing small amounts of solvent is shown in Figure 2.18. In this example, solution is placed into a flask fitted with a rubber stopper containing a glass tube about 6–8 cm in length. The glass tube is connected to a diaphragm pump

13. A pump capable of 45 Torr and 17 L/min works well; Model N726FTP. KNF Neuberger, Inc., Trenton, NJ, -690-890-8600.

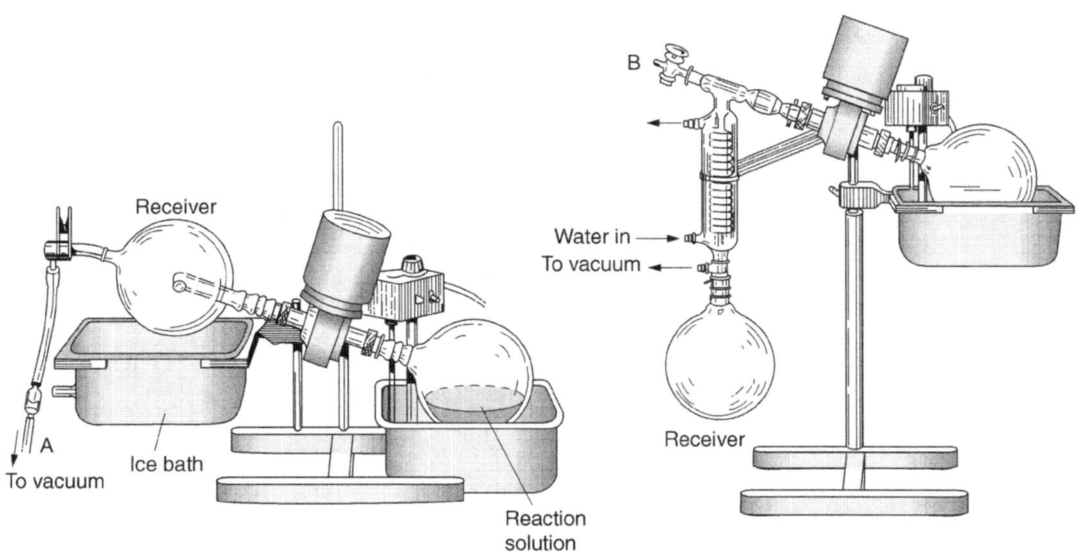

Figure 2.17   Rotary evaporators

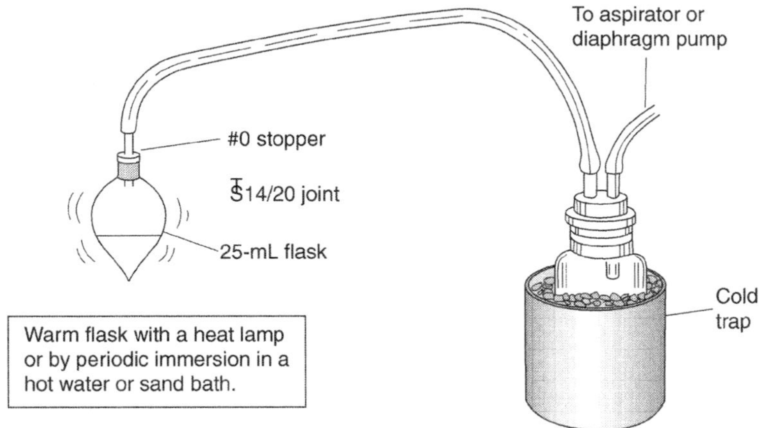

Figure 2.18   Simplified method for small-scale solvent removal

or an aspirator (through a trap such as the ones shown in Figure 5.5) and swirled by hand over a steam bath or in front of a heat lamp. The evaporation of many solvents with boiling points below 100°C takes only a few minutes with this procedure. This simple method can be conveniently used for flask sizes from 5 mL to 100 mL (solution volumes from 2 mL to 50 mL).

18. What are the advantages of using a rotary evaporator to remove solvent rather than a simple distillation apparatus?

2.6 Connections

Glass Tubing

Because bending, cutting, and handling glass tubing represent a potential source of thermal burns and cuts, proper procedures are important. Soft-glass tubing can be conveniently worked with an ordinary Bunsen burner, but if more elaborate glassblowing operations are required, such as joining two glass tubes, Pyrex tubing heated by an oxy–gas torch is recommended because it is much easier to control the softness of Pyrex glass and produce a strain-free joint.

Tubing with a diameter of 1 cm or less can be conveniently cut by scratching the tube with a sharp triangular file, carborundum, or diamond cutter; wetting the scratch; and pulling while applying firm pressure away from the scratched side (Figure 2.19a and b). A piece of cloth placed over the tubing will help to protect both hands. The raw (sharp) ends of all glass tubes should be fire polished by rotating the cut surface in a burner flame until the edge becomes rounded (Figure 2.19c).

Tubing can be sealed by continuing to rotate the end of the tube in the flame until the soft edge pulls together. When the tube is entirely closed, remove it from the flame and without delay blow gently *into the other end* to make the thickness of the wall even at the closed end. A small glass bulb can be made by gathering a blob of red-hot glass, removing it from the flame, and blowing firmly while the tube is rotated. Care must be taken not to produce thin areas in the wall of the bulb.

To bend a piece of soft-glass tubing, simply move the tube lengthwise in the flame, rotating the tubing as you do so. Care must be taken to heat the tube evenly and not to allow the glass to become soft enough to sag under its own weight. A burner equipped with a wing tip is best for this work (Figure 2.20a). The hot tube should be removed from the flame and then bent to the desired shape (Figure 2.20b). If several bends are to be made in succession and all must be in the same plane, carry out the bending and press the hot, pliable tubing against a piece of asbestos board.[14]

14. Transite, which is a mixture of asbestos and concrete that is safer to handle than pure asbestos, is often used.

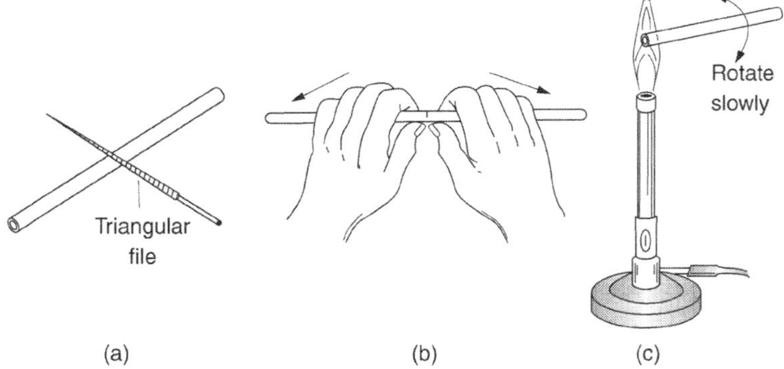

Figure 2.19 Cutting glass tubing: (a) scoring glass tubing; (b) breaking the tube (for added safety, cover tube with cloth); (c) fire polishing

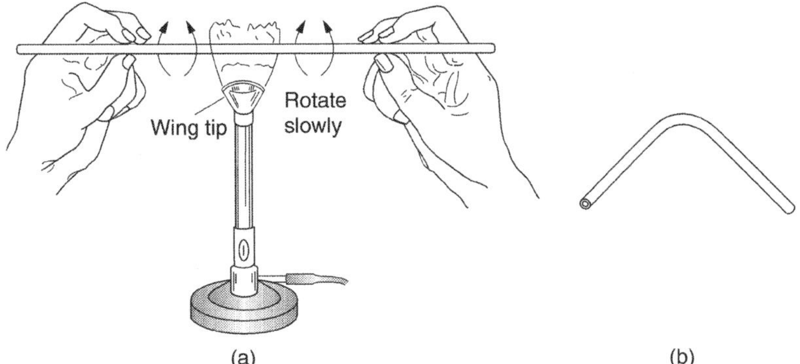

Figure 2.20 Bending glass tubing: (a) burner with wing tip for even heating; (b) right-angle bend

Cork and Rubber Stoppers

Although ground-glass fittings (to be discussed) are extremely common, there are occasions when corks and rubber stoppers are convenient and appropriate to use. To bore a hole in a cork, first soften the cork either by rolling it on a lab bench with a wood block or by using a cork softener (Figure 2.21a). Choose a sharp cork borer (Figure 2.21b or c) with a slightly smaller hole than desired and bore the hole by twisting and pressing the borer against the face of the cork. The cork should be placed against a soft wood or cardboard surface throughout the process. This will avoid chunks of cork breaking off as the borer emerges from the opposite face. Do not ruin the cutting edge of the borer by forcing it against any hard surface. Holes can be bored in a rubber stopper by a similar technique, but a very sharp cutting edge on the

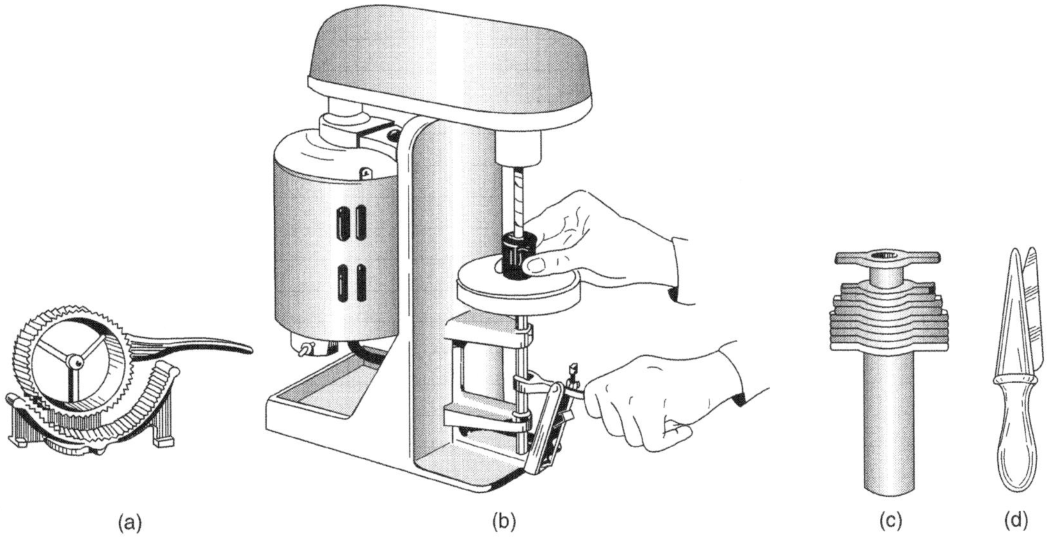

Figure 2.21 Equipment for boring cork and rubber stoppers: (a) cork softener; (b) powered cork borer; (c) manual cork borer; (d) cork-borer sharpener

borer and a lubricant, such as glycerol or silicone oil, are essential. A tool for sharpening a brass borer is shown in Figure 2.21d.

Because rubber stoppers often swell in contact with organic liquids and vapors, they are often less desirable than ground-glass joints when a tight fit is needed, or corks when a tight fit is unnecessary. If rubber stoppers must be used, those made of neoprene rubber are softer and more resistant to heat, oxidation, solvents, and other chemicals than are ordinary rubber stoppers.

To insert a glass tube or thermometer into a cork or rubber stopper:
- Lightly apply a lubricant such as glycerol, silicone oil, or stopcock grease to the tube.
- Grasp it close to the stopper with a towel, and carefully twist the tube through the hole with firm, steady pressure.

Forcing glass tubes through holes in a stopper is a mistake and too often results in serious cuts from broken glass. *Do not try to force a tube through an opening that is too small.*

To remove a thermometer or glass tube that is stuck in a stopper, place a slightly larger cork borer around the tube and work it through the stopper until the thermometer or glass tube is free.

Flexible Tubing

Connections between glass tubes are sometimes made with rubber or plastic tubing. In general, thin-walled rubber tubing should only be used for water hoses to condensers or natural gas hoses to burners, while the heavy-walled tubing is normally used for vacuum hoses. All types of rubber tubing deteriorate with time and must be carefully and regularly inspected and replaced.

Soft Tygon[TM] tubing is more chemically resistant than rubber, is transparent, and generally lasts much longer before it needs to be replaced. Unfortunately, the flexibility of this plastic is the result of a plasticizer (such as diethyl phthalate, an ester), which can be extracted from the plastic if it comes into contact with an organic solvent. For that reason, flexible joints between glass tubes that will carry organic liquids must be kept very short.

Ground-Glass Joints

Standard-taper joints (Figure 2.22a) are extensively used in modern chemical apparatus. They have the advantages of allowing rapid assembly and disassembly, and of providing a durable gas- and vacuum-tight seal. Though the use of ground-glass joints eliminates the type of contamination that can result from the use of cork and rubber stoppers, the lubricant applied to the joint (see below) does dissolve in many organic solvents and can become an undesired contaminant in a reaction mixture. Although the ball-and-socket type (Figure 2.22b) is frequently used where flexibility in the joint is required, the straight standard taper is the type most often encountered.[15] Because of the rigidity of a standard-taper joint, care must be taken to avoid undue strain on the apparatus when it is clamped to a solid support such as a ring stand.

A good grade of stopcock grease should always be applied lightly to the ground-glass surface of a joint to provide a tight seal, and to prevent the joint from "freezing."

15. The size of a straight, tapered, interchangeable joint is designated as 14/20, 19/22, 24/40, and so on, and refers to the diameter of the large end of the joint (mm) and the length of the taper, respectively. The sizes of ball-and-socket joints are similarly designated 18/9, 28/12, and so on, and refer to the diameter of the ball and the inside diameter of the glass tube.

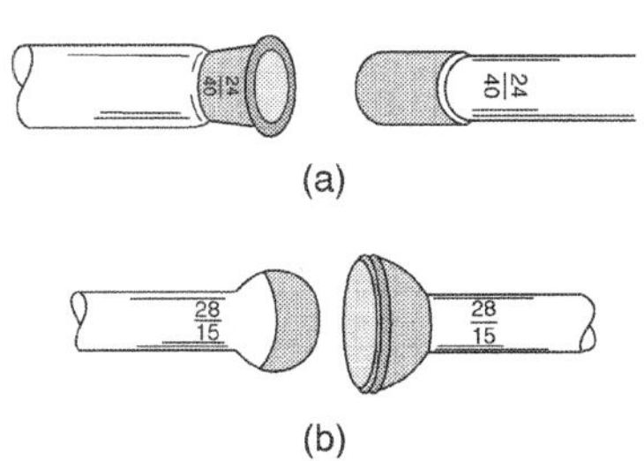

(a)

(b)

Figure 2.22 Ground-glass joints:
(a) standard taper; (b) ball and socket

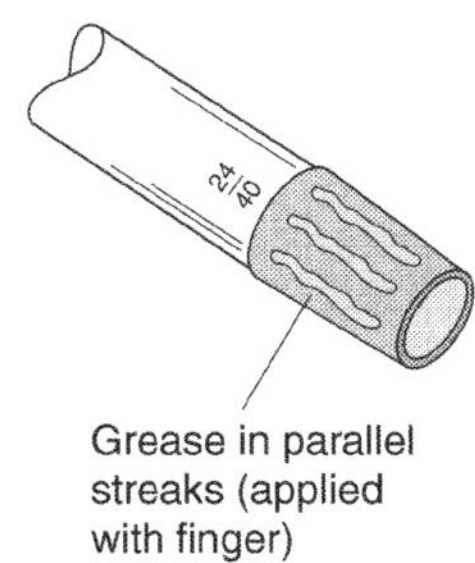

Figure 2.23 Proper method for
greasing a standard-taper joint

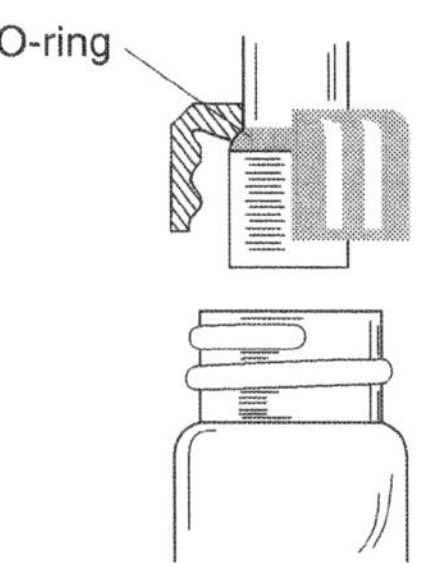

Figure 2.24 Greaseless tapered
joint with threaded cap and O-ring

The grease should be applied in four or five streaks that run lengthwise with the male joint and are placed evenly around the taper (Figure 2.23). The joints are then pressed firmly together without twisting (or with only slight twisting motion) until the grease streaks spread and meet. Excess grease serves no useful purpose, and as was already mentioned, can contaminate the reaction mixture. After each use, the joint should be wiped free of grease and cleaned with methylene chloride or diethyl ether.

If a joint becomes stuck (frozen), it may sometimes be freed by tapping it with the wooden handle of a spatula. As a last resort the outer joint can be *gently* and evenly heated with a torch (solid blue flame) until it expands and breaks away from the inner male section.[16]

A greaseless connection that uses a standard taper available in several small sizes (e.g., 14/10 and 7/10) and a threaded cap with an O-ring to make the final seal is shown in Figure 2.24. Because of the design, the connection is gas and vacuum tight, but the content of the flask does not make contact with the O-ring. This type of connection is expensive but widely used in microscale apparatus.

16. The apparatus must be clean and dry before attempting this procedure.

19. Explain the procedure and proper precautions associated with:
   (a)   inserting a glass tube through a hole in a rubber stopper
   (b)   separating a frozen standard-taper joint
   (c)   cutting a piece of 7-mm-diameter glass tubing

20. Name two appropriate solvents for the removal of stopcock grease from ground-glass joints.

2.7 Inspection and Care of Laboratory Apparatus

Before assembling any glassware, a careful check for flaws such as cracks or chips should be made. Electrical equipment should be checked for frayed cords or ungrounded plugs. Apparatus clamped to a support rack should be attached firmly, but without putting strain on any glassware. For macroscale reactions, apparatus should be clamped sufficiently far above the bench that a cold bath could be used quickly to moderate a reaction that has become too vigorous. Before starting any reaction, check all glass joints, stoppers, and hose connections for a firm fit, and check the alignment of stirring motors.

Once a reaction is over, glassware should be disassembled and cleaned at the earliest opportunity. Not only does this reduce clutter, but joints are less likely to become stuck, and potentially toxic residues can be properly eliminated or contained before they can do any harm. The most useful cleaning agents for glassware are the various laboratory detergents together with a generous application of "elbow grease." Brushes and steel wool are usually sufficient for removing hardened residues and tars, but if these are inaccessible it may be advisable to soak the residue with acetone or methylene chloride. Most laboratory greases are soluble in methylene chloride or diethyl ether. It is not usually fruitful or advisable to use strongly acidic cleaning solutions or strong oxidizing agents to remove organic residues.[17] Nitric acid particularly must be avoided, due to the possibility of explosive compounds being formed. However, hot solutions of sulfuric and nitric acids can be used for removal of carbonaceous residues.

Glassware cleaned with detergent should be thoroughly rinsed, allowed to drain, and then dried either in an oven or at room temperature overnight. If the piece of glassware is needed for immediate use, it can be rinsed with a little acetone and dried by placing a suction hose from an aspirator or vacuum line into the vessel and drawing air through it for a few minutes. (Compressed air is not as suitable for drying, because it is frequently contaminated with dirt particles and droplets of water or oil.)

17. A hot alcoholic solution of potassium hydroxide is sometimes useful for removing residues (especially silicone lubricants), but can etch the surface of the glass.

CHEMISTRY 203

CHEMISTRY 203
LABORATORY SCHEDULE

WEEK	EXP #	TITLE
1	EXP. 2	Orientation/Determination of Melting Points
2	EXP. 3	Thin Layer Chromatography of analgesics
3	EXP. 7	Caffeine: Isolation of Alkaloid
4	EXP. *	Gas Chromatography
5	EXP. 12	Fractional Distillation: GC
6	EXP. *	Chemistry of Hydrocarbons
7	EXP. 10	Extraction and Crystallization: Acid – Base Properties
8	EXP. 22	Triphenylcarbinol Addition of a Grignard Reagent to a Ketone
9	EXP. 19	Introduction to Infrared Spectroscopy
10	EXP. 31	Introduction to Nuclear Magnetic Spectroscopy
11	EXP. 31	NMR Spectral Analysis
12	EXP. 36	Carbocation Rearrangements: Benzopinacolone
13	EXP. 16	Methylcyclohexenes: Alcohol Dehydration
14	EXP. 39	Hydoboration – Oxidation of 1 – Hexene

LABORATORY MAKE UP AT THE DISCRETION OF THE LABORATORY INSTRUCTOR.

* Original Lab

CHAPTER 3

Examination of Pure Compounds

3.1 Melting Points

The **melting point** (or **freezing point**) of a pure compound is defined as *that temperature at which the solid and liquid phases are in equilibrium at some particular pressure, usually taken as* 1 *atm or* 760 *mm Hg.*

$$\text{Solid } Q \xrightleftharpoons{T^{\circ}=mp} \text{Liquid } Q$$

The definition implies that when solid Q is in heterogeneous equilibrium with liquid Q at some temperature, the melting point (mp), a slight increase in the temperature will result in the complete conversion of solid Q to the liquid phase. Conversely, a slight decrease in temperature will result in the complete conversion of liquid Q to solid Q.

Because the volume change on going from the solid phase to the liquid phase is usually very small, a change in pressure has very little effect on the melting point. For example, an increase in pressure from 1 to 2 atm lowers the melting point of ice by only 0.0024°C, an amount too small to be of significance in routine organic laboratory work.[1]

Experimental Methods

The most commonly used technique for determining melting points is to place a very small amount of sample in a thin-walled capillary tube (*ca.* 1.5 mm o.d.) sealed at one end, place the tube next to a thermometer in a metal block or oil bath, gradually increase the temperature of the bath, and note the temperature range over which the material melts.

After a few crystals of the compound have been introduced into the open end of the capillary, the latter should be tapped or dropped several times through a glass tube at least 1 ft in length onto the table top to pack the compound into the capillary tube.

One simple melting-point apparatus shown in Figure 3.1a employs an electrically heated metal block into which the melting-point capillary and thermometer are inserted.[2] A light and a magnifier allow the sample to be viewed as the block is heated. (Several melting-point capillaries can be inserted at one time for comparisons.)

1. Whether a melting point is increased or decreased by an increase in pressure depends only on the relative sign of the volume change on going from the solid to the liquid phase.

2. MelTemp Apparatus; Laboratory Devices, P.O. Box 6402, Holliston, MA 01746-6402.

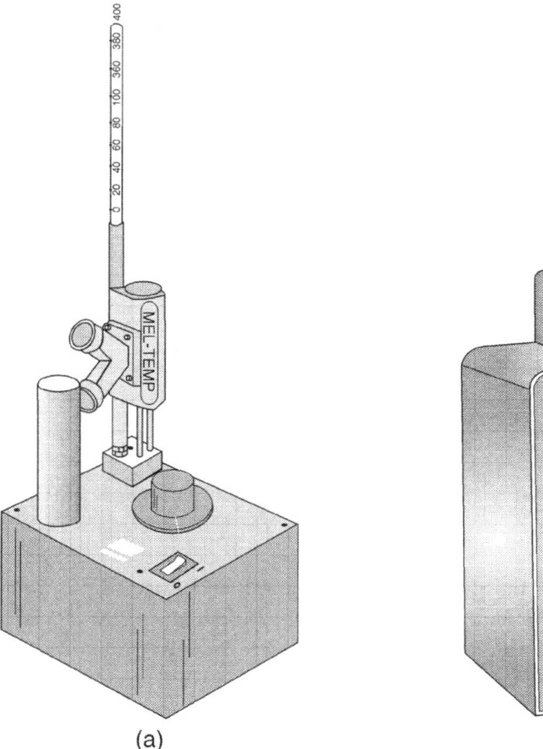

Figure 3.1 Capillary melting-point apparatus: (a) heated block with magnifier and light; (b) heated oil bath with stirring device, magnifier, and light

Figure 3.1b shows a more sophisticated apparatus that employs a stirred silicone oil bath and a periscope so the mercury (or dyed alcohol) thread in the thermometer and the crystals in the capillary tube can be viewed simultaneously during heating. One essential feature of each apparatus is that the thermometer and capillary are close together so that they experience the same temperature.

The apparatus can be heated rapidly to within about 15°C of the expected melting point, at which time the rate of heating should be adjusted to cause only a 1–2° rise in temperature per minute. Keep in mind that there can be an appreciable lag in the response of the apparatus to any change in the voltage applied to the heater circuit. The temperature at which the sample *first* begins to melt and the temperature at which it is *totally* melted comprise the **melting range**. Although the melting point is the highest temperature in this range, *the terms melting point and melting range are often used interchangeably*. If the temperature of the apparatus is increased too rapidly, it will overshoot the correct melting point because the temperature within the capillary will not have had an opportunity to attain thermal equilibrium with the heated metal block or oil bath. Not only will the recorded melting point be incorrect, but the melting range will appear to be much smaller than it actually is.

For extremely small samples or for greater accuracy in determining a melting range, it is advantageous to use a microscope with a hot stage (Figure 3.2a). The sample is placed between two microscope slides on an electrically heated metal block and carefully observed through the microscope during the melting process. A less expensive heated-stage apparatus is shown in Figure 3.2b.

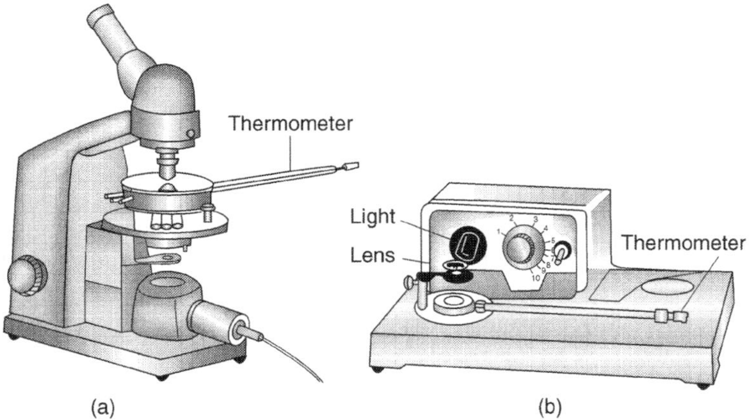

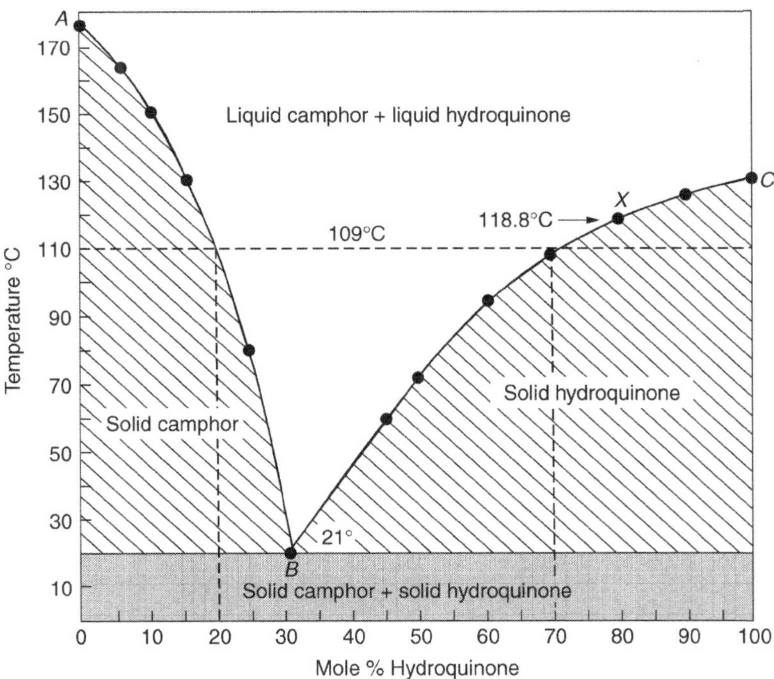

Figure 3.2 Melting-point apparatus that use a heated stage: (a) microscope with hot stage; (b) heated stage with viewing lens and light

Phase Behavior

Some understanding of the physical nature of the melting process is essential before an explanation can be given of the use of melting points as a method of identification and a criterion for purity. It is at the melting point of a pure crystalline compound that sufficient thermal energy has been introduced to begin breaking the crystal lattice. For many (but not all) pure compounds the range of temperature change, or melting range, over which this breakdown occurs will be from 0.5°C to 2°C. The melting range, therefore, is sometimes taken as circumstantial evidence of purity.

The anticipated melting (or freezing) behavior of a mixture of compounds can be conveniently explored by examining a **temperature–composition phase diagram** such as the one illustrated in Figure 3.3 for hydroquinone and camphor. The left and

Figure 3.3 Temperature–composition phase diagram for a mixture of camphor and hydroquinone

23

right ordinates (vertical axes) of the plot represent pure camphor and hydroquinone with melting points of 178°C and 130.8°C, respectively. The region of the graph above and in between curves AB and BC represents a liquid solution of hydroquinone and camphor, while the area to the right of curve BC is pure solid hydroquinone in contact with the liquid solution. The area to the left of AB is pure solid camphor in contact with the solution. A mixture of solid compounds exists below the line drawn through point B. Line AB represents an equilibrium between pure solid camphor and the liquid solution. For example, at a temperature of 109°C, two equilibria are possible. Either pure solid camphor can be in equilibrium with a liquid solution of about 80 mole % camphor and 20 mole % hydroquinone or pure solid hydroquinone can be in equilibrium with a liquid solution of about 70 mole % hydroquinone and 30 mole % camphor.

It is instructive to consider the fate of a liquid solution of a certain composition as it is cooled. Consider a liquid solution of 20 mole % hydroquinone and 80 mole % camphor at a temperature of 120°C. No change will occur until the solution is cooled to 109°C, at which time pure camphor crystallizes. A further drop in temperature causes more camphor to crystallize and the remaining solution becomes richer in hydroquinone. We are now proceeding down and to the right along line AB until we reach a temperature of 21°C. This point on the curve (point B on the diagram) represents the **eutectic temperature** and the composition is known as the **eutectic composition**. This is the lowest temperature at which a liquid solution can exist in equilibrium with both of the pure solid components. If the temperature is lowered below 21°C, all of the remaining liquid will solidify and the overall composition of the solid will be exactly 20 mole % hydroquinone and 80 mole % camphor (the starting composition). In a similar manner, we could have started with a liquid solution rich in hydroquinone, cooled it until pure hydroquinone began to crystallize, and then followed line BC to the left and down until the same eutectic point was reached.

If we reverse the foregoing exercise and begin with a solid mixture of 20 mole % hydroquinone and 80 mole % camphor, melting begins at the eutectic temperature and continues over a wide range, following the path already described but in reverse. The material that melts first has the eutectic composition and melting continues until all of the hydroquinone is in the liquid phase. Then the temperature begins to rise as the remaining pure camphor continues to melt as we proceed up and to the left along line AB. When the temperature exceeds 109°C, all of the starting solid will have melted. In actual practice a solid mixture that contains only a few percent of a second component may exhibit a melting range of only a few degrees, because the experimental technique does not allow detection of the very small amount of melting that begins at the eutectic temperature. *Material having the eutectic composition will, of course, always exhibit a sharp melting point, which could mislead you into thinking you have a pure compound.*

Although the previous description of a type of melting behavior applies to a large number of binary mixtures, two other types of behavior are also commonly encountered. The first, illustrated by the temperature–composition diagrams in Figure 3.4, illustrates cases in which **compound formation** takes places between the two components. The compound will usually have a simple composition designated as AB, A_2B, A_2B_3, and so on, and can have a melting point associated with the compound that is either above or below that for either pure component. There are now two eutectic points, and mixtures of A and B will exhibit a wide melting range.

In still another type of phase behavior, components A and B can form a **solid solution** in which the composition of each crystal consists of a random mixture of A and B. This is in contrast to the crystals of a compound such as AB or AB_3, in which the composition of each crystal is the same. The melting point of a solid solution can be slightly above, slightly below, or between those for the pure components; the

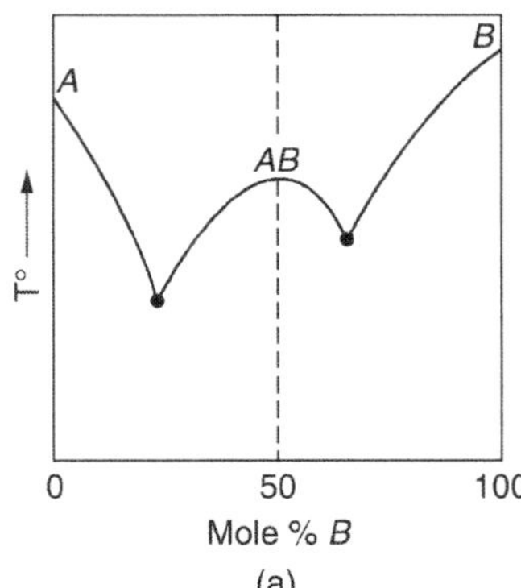
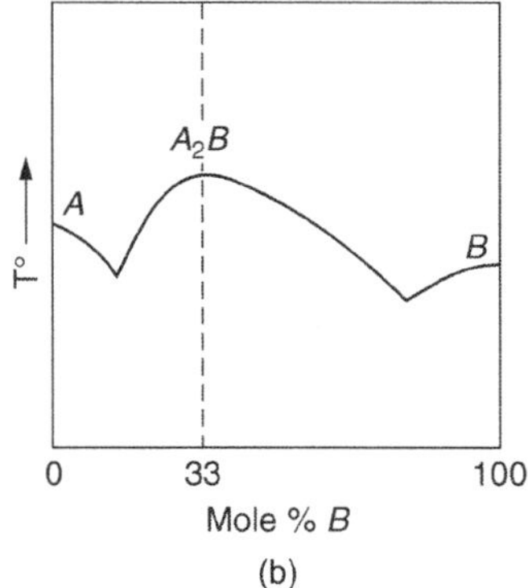

Figure 3.4 Temperature–composition phase diagrams for components that form simple compounds

melting range can be rather narrow. An example of the temperature–composition diagram for a solid solution is shown in Figure 3.5.

Method of Mixed Melting Points

Subject to the considerations discussed previously, melting points can sometimes be used to help determine whether two compounds with the same melting point are the same or different. The two compounds are intimately mixed by grinding them together in the depression of a spot plate with a stirring rod, and a melting point is then determined.[3] As demonstrated in the discussion on phase behavior, the melting point of a mixture of two different compounds varies from that of either of the compounds unless they form a solid solution. Therefore, if the melting point of the mixture is different from that of either compound,[4] the two compounds are *definitely different*. However, if the melting point of the mixture is the same as that of the two pure compounds, *the compounds are either the same or they form a solid solution.*

3. It is preferable to carry out the mp determination on the mixture and at least one of the pure compounds simultaneously in the same apparatus to ensure that the experimental conditions for the two determinations are identical.
4. This means that the melting point can be sharp but different or have a wide range and be different.

Determination of Melting Point

A. Determination of Melting Point

Background Reading
Section 3.1 in the laboratory text

Timing
(1.5 h)

Determine the melting range of two known compounds using an apparatus such as the Mel-Temp unit pictured in Figure 3.1a. Samples with convenient melting points might include 9-fluorenone (listed mp 83°C) and benzoic acid (listed mp 122°C).

Remember that the key to a good melting-point determination is maintaining a slow increase in temperature (1–2°/min) when you are within 10° of the melting point. For this reason, if the expected melting point is not known, it is best first to obtain a very approximate melting point by increasing the temperature rather quickly and then to determine it more exactly with a second sample by slow heating in the vicinity of the approximate melting point.

Try mixing 10–20% (estimate by eye) of one of the previous samples for melting-point determination with the other sample (with a small spatula on a piece of glazed paper or a watch glass) and obtain a melting range on the mixture. How does it compare with the melting range for each pure sample?

EXERCISES

1. Describe how mixed melting points can be used to help identify an unknown compound.
2. When compounds with different melting points are mixed, is it possible to obtain a narrow melting range for the mixture? Explain.

B. Unknown Identification

Obtain an unknown sample and a list of possible compounds from your laboratory instructor. Carefully determine the melting point and check your identification by running a mixed melting point with an authentic sample.

<u>**Determination of Melting Point**</u>

1. Read the booklet and learn how to operate the computerized melting point apparatus.

2. Find out the melting point of the compounds we will be using ,use CRC Handbook.

3. Read the attached handout , understand the procedure ,use the melting point apparatus to get the melting point of these compounds . Record data in the following table:

Data Table

Compound	M.P.(CRC)	Experimental M.P.	Range
Salicylic acid			
P-dichlorobenzene			
Benzamide			
Acetamide			
Acetanilide			
Known Mixture			
Unknown # ————			
Unknown with known to check			

* Unknown # ——— is ————————————

Thin-Layer Chromatography

Because of its simplicity and rapidity, thin-layer chromatography (TLC) has been widely adopted for the qualitative examination of mixtures of nonvolatile compounds. Specific uses include monitoring the purification of a sample, quickly establishing optimum conditions for carrying out a synthesis, examining crude reaction mixtures, determining the homogeneity of a sample, and, in conjunction with other evidence, establishing whether or not samples from different sources are identical. The thin-layer method can be used for qualitative analysis of quantities as small as 10 micrograms.

Adsorbents used for TLC are very finely divided with a typical particle diameter of *ca.* 40 μm. The adsorbent is normally mixed with a **binder** such as hydrated calcium sulfate (plaster of paris) to hold the layer on its support (an aluminum, glass, plastic, or fiberglass sheet or plate). Most thin-layer separations are carried out on silica gel or aluminum oxide, but other supports are available. The thickness of the coating

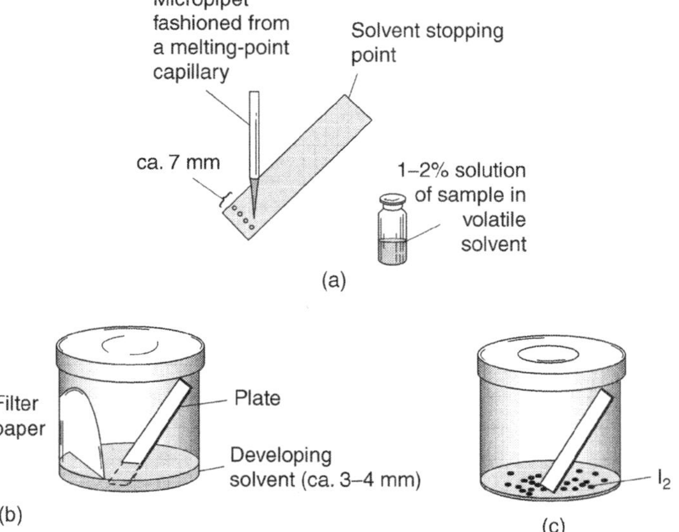

Figure 4.5 Thin-layer chromatography: (a) spotting mixture; (b) developing chromatogram; (c) visualizing chromatogram

used for analytical separations is about 0.25 mm. Although thin-layer plates are commercially available, a simple and inexpensive process for coating microscope slides has been described.[9]

In the following paragraphs and Figure 4.5 is a procedure for doing TLC on small, commercial rectangular plates (*ca.* 2 or 3 cm $\times$ 9 cm).

9. J. J. Peifer, *Mikrochim. Acta*, **3**, 529 (1962); other procedures that use alumina, florisil, and cellulose powder are described in the article.

- **Spotting.** Several spots of the dissolved sample are applied about 7 mm from the bottom of the plate and spaced about 5 mm apart. This procedure requires only a few mg of the sample and is best done by using a few μL of a 1–2% solution of the sample in a volatile solvent (dichloromethane, ethyl acetate, etc.). Spotting is readily achieved by using a micropipet made by drawing out a melting-point capillary at the center in a microburner and breaking it into two parts. Such a device fills by capillary action. The tip of the pipet is then touched quickly, once or twice, to the same location on the plate to achieve a concentrated spot with small dimensions (Figure 4.5a).[10]

- **Developing.** The plate is placed (spotted end down) in a chamber containing a 4–5-mm depth of the developing solvent[11] and having an atmosphere saturated in solvent vapor. Solvent moves up the plate by capillary action. Development is complete in 5–7 min when the solvent reaches the top of the plate and can be carried out in a screw-cap jar (Figure 4.5b). The filter paper promotes rapid equilibration of solvent liquid and vapor in the chamber. (In many thin-layer chromatographic separations, resolution can be improved by allowing the plate to dry and developing it a second time, provided that the spots have traveled less than half the distance that the solvent front has traveled.)

- **Visualizing.** If the compounds being chromatographed are colorless, various procedures for visualization can be used after the thin-layer plate has been removed from the developing chamber and allowed to stand for a few minutes. A rather useful (and for many compounds, nondestructive) method is to put the microscope slide or plate in a chamber containing iodine (Figure 4.5c).[12,13] Organic materials readily absorb the iodine vapor to produce brown spots on

10. It is important not to gouge a large hole in the adsorbent at the point of application, because this will result in spots (after development) with very distorted shapes that can be difficult to interpret. If the pipet is allowed to contact the plate for more than a fraction of a second, the amount of sample applied to the plate can be too large, which will result in a significant loss in resolution.

11. The solvent level *must* be below the level of the spots on the plate.

12. This method is not particularly effective when the silica gel G contains a fluorescent dye.

13. An effective alternate procedure is to sprinkle the developed plate with a mixture of 10% iodine on TLC-grade silica gel and then brush it off.

Table 4.5 Solvent Systems for TLC on Silica Gel

Class of Compounds	Solvents [a] (by volume)
Alcohols (ROH)	(a) Cy:EA (1:1 or 1:2)
	(b) P:DE (10:1)
Amides (RCONH$_2$)	EA:M (5:1)
Amines (RNH$_2$), amino alcohols	(a) M:Ch (2:3) 1% (by vol.) of 33% NH$_4$OH
	(b) EA:M (1:1)
Ammonium salts, quaternary (R$_4$N$^+$X$^-$)	M:Ch:conc. HCl (6:6:1)
Carboxylic acids (RCO$_2$H)	Ch (saturate with 90% formic acid)
Enolic diketones (—COCH$_2$CO—)	(a) EA:M (4:1)
	(b) EA
Esters (RCO$_2$R)	Cy:EA (1:1 or 1:2)
Hydrocarbons (RH)	(a) P
	(b) Cy
	(c) B:DE (2:1)
	(d) Cy:EA (3:1)
Ketones (RCOR)	(a) Cy:EA (1:1 or 1:2)
	(b) DE:PE (1:2)
	(c) DE:H (1:9)
Lactones (cyclic esters)	(a) Cy:EA (1:1 or 1:2)
	(b) DE:PE (1:2)
	(c) EA
Phosphonium salts (PR$_4^+$X$^-$)	M:EA (1:1)
Phosphorus ylides (R$_3$P$^+$CH$_2^-$)	M:EA (1:1)

[a] B = benzene, Ch = chloroform, Cy = cyclohexane, DE = diethyl ether, EA = ethyl acetate, H = hexane, M = methanol, P = pentane, PE = petroleum ether

a white to light-tan background. Plates treated in this manner will fade over a period of about 1 h but can be retreated if necessary.

In a second widely used procedure a fluorescent dye is incorporated with the adsorbent. Irradiation of the developed plate or slide causes the fluorescent dye to glow, and if the spots are compounds that absorb the emitted light, they will appear dark on a bright background.

Many reagents, some more specific than others, can be carefully sprayed on the developed plate to produce colored spots when certain types of compounds are present that react with the reagent in the spray.[14]

Because the choice of the developing solvent depends on experience, a summary of some useful solvent systems for TLC on silica gel and alumina, respectively, is listed in Tables 4.5 and 4.6. A rapid procedure for testing the effectiveness of different solvent systems consists of spotting the sample on a thin-layer plate, filling a capillary pipet with the solvent to be tested, and holding the tip of the capillary on the center of the spot so that the solvent creates a circular area approximately 1.0–1.5 cm in diameter and partially resolves the mixture (Figure 4.6). For the solvent to be effective, the components should travel about $\frac{1}{3}$ to $\frac{1}{2}$ the distance traveled by the solvent front (Figure 4.6a). Thus, Figures 4.6b and 4.6c represent unsuitable solvent systems.

14. A list of more than 200 reagents can be found in J. G. Kirchner, *Thin-layer Chromatography*, 2nd ed., Wiley–Interscience, New York, 1978, Ch VII.

Class of Compounds	Solvents[a] (by volume)
Alcohols (ROH)	Cy:EA (1:1 or 1:2)
Hydrocarbons (RH)	P
Ketones (RCOR)	Cy:EA (1:1 or 1:2)
Lactones (cyclic esters)	Cy:EA (1:1 or 1:2)

[a] See footnote for Table 4.5.

Table 4.6 Solvent Systems for TLC on Alumina

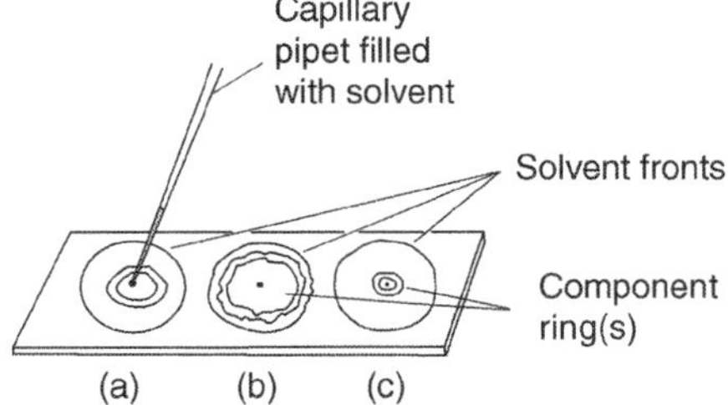

Figure 4.6 Method for determining a good solvent system: (a) good development; (b) and (c) poor development

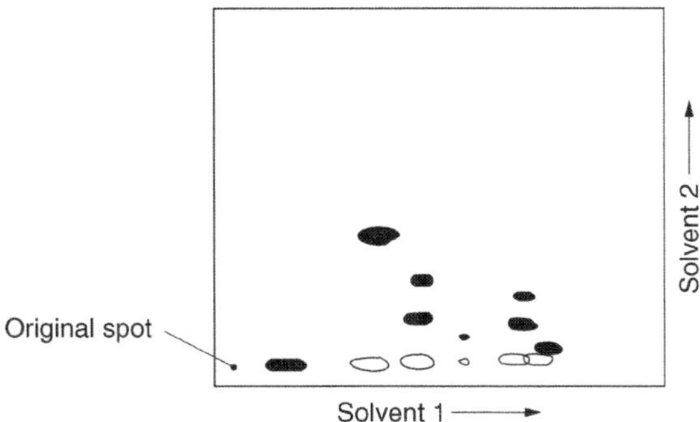

Figure 4.7 Two-dimensional TLC

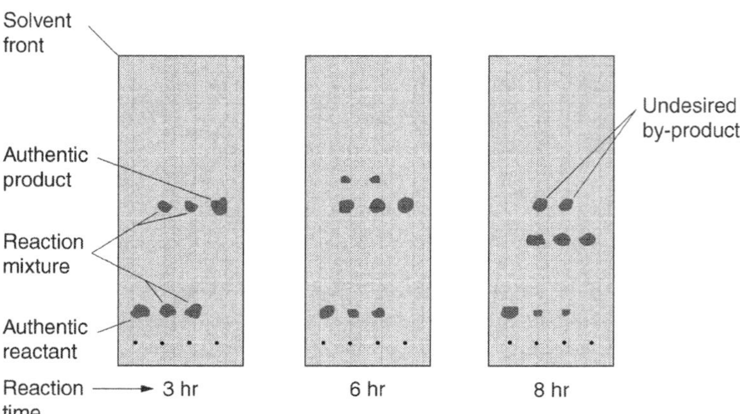

Figure 4.8 Thin-layer chromatograms of a reaction mixture

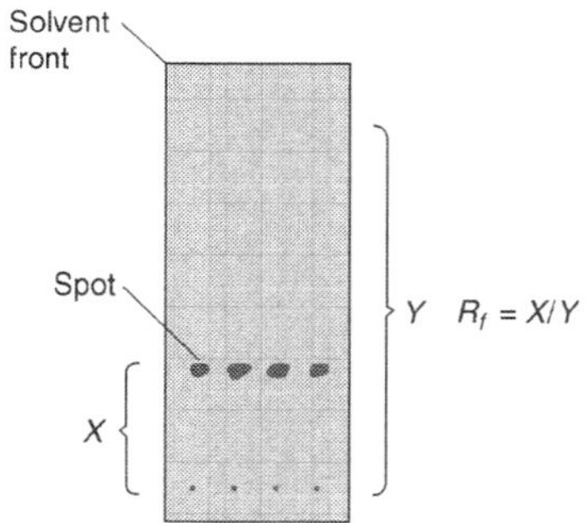

Figure 4.9 The R_f value

The use of longer TLC plates will require a longer time for development, but has the advantage that more components can be clearly separated because of the greater distance traveled by the solvent front. For square plates, **two-dimensional chromatograms** can be obtained. Thus, if partial separation is achieved with one solvent system, the plate can be dried, rotated 90°, and developed in a second direction with a different solvent system as illustrated in Figure 4.7.

At this point it might be useful to consider how TLC can be used to monitor a reaction. Figure 4.8 shows three thin-layer plates of a reaction mixture taken after 3, 6, and 8 h of reaction time. The chromatograms shown in the figure have been simplified. Chromatograms of crude reaction mixtures would exhibit many more spots due to various reagents, side-reaction products, and impurities, in addition to the product and starting compounds. Note that authentic starting material and product should be included on the plate for comparison. It can be seen that at the end of the 8-h period most of the starting material has been exhausted, but a substantial amount of an undesired by-product has been formed. Apparently, a time of approximately 6 h would maximize the yield of desired product while avoiding most of the by-product.

To compare thin-layer chromatograms of two substances suspected of being identical, we normally define an R_f value as the distance moved by the spot relative to that moved by the solvent front (Figure 4.9).

Recognizing that the R_f value depends on the temperature, layer thickness, moisture content of the adsorbent, degree of solvent saturation of the atmosphere in the development chamber, and sample size, in addition to the type of adsorbent and solvent used, it is essential that comparisons be made with the unknown and known samples, side by side on the same plate. Furthermore, several different solvent systems should be used, and chromatograms consisting of a mixture of the unknown and known materials should be run. When mixed, two materials that have nearly identical R_f values will often appear as an exceptionally elongated spot or a partially overlapped pair of spots, which would clearly indicate that the substances were not identical.

As with other chromatographic techniques, tentative identifications by TLC should always be confirmed by other, more conclusive, procedures such as spectroscopy. To obtain a sample for spectroscopic examination, it may be necessary to use a thicker adsorbent layer (e.g., 1–2 mm), which will allow hundreds of milligrams of sample to be chromatographed.[15] The sample is usually applied as a streak across the plate near one end rather than as a spot. If the resolved components are not visible, either a nondestructive visualization technique, such as ultraviolet light or possibly iodine vapor, must be used or a destructive technique, such as a specific spray reagent, must be applied to a narrow band at one edge of the plate. When the locations of the components have been established, they can be scraped from the surface of the plate and eluted from the adsorbent with a polar, volatile solvent.[16]

15. Layers of 2–5 mm can be used to separate quantities in excess of 500 mg.

Experiment 3

Thin Layer Chromatography

Instructions:

Spotting : Dissolve a small amount of each known compound (*Naphthalene, Vanillin, and Allylurea)* in 1ml of methylene choloride in a small vial. Do spotting using capillary tubes.

Developing: Develop with Ethylacetate.

Visualizing: Use Iodine chamber.

(For more details about the above techniques see page 61 & 62)

Procedure:

I. Run a TLC for each of the three substances provided, these are :

  Naphthalene, Vanillin, and Allylurea.

II. Run a TLC for one of the following mixtures:
   a) Vanillin and Allylurea
   b) Naphthalene and Vanillin
   c) Naphthalene and Allylurea

III. Run a TLC for one of the unknowns provided.

Write a complete report, including the answers to the following questions:

1. What conclusion can you draw about the relative polarity of the three authentic samples?
2. What additional methods might be used to make the spots on a thin-layer chromatogram visible?
3. Two materials on a thin-layer chromatogram follow the solvent front (Rf = 1). Does this show that they are identical? What additional experiments are necessary?

Experiment 3

Data :

Substance	Rf = x / y
Naphthalene	
Vanillin	
Allylurea	
Known Mixture	....
Unknown #	

Unknown # ....   is ....

Explain why ? ....

....

Caffeine; Isolation of an Alkaloid

Tea and coffee are popular beverages known from antiquity. Both are rich in caffeine, a central nervous system (CNS), respiratory, and cardiac stimulant. Caffeine is also a diuretic agent (tending to increase urine production) and can produce nervousness, insomnia, and racing heartbeat (tachycardia). Heavy consumption can lead to a caffeine dependence such that headache and even nausea can be experienced without it.

Caffeine is one of the **alkaloids**, which are basic nitrogen-containing compounds found in certain plant materials. Structurally, it is 1,3,7-trimethylxanthine. It is also a derivative of purine and is therefore related to the purine bases in nucleic acids and to uric acid. Two closely related compounds are the dimethylxanthines, theophylline, which occurs along with caffeine in tea, and theobromine, which occurs instead of caffeine in cocoa.

Xanthine   Purine   Uric acid

EXERCISES

1. Draw the structures of theophylline (1,3-dimethylxanthine) and theobromine (3,7-dimethylxanthine) and compare them to that of caffeine.

2. Why is purine an alkaloid?

The caffeine content of coffee beans is about 1–2% by weight and in tea leaves is about 2–4%. Because of differences in methods of brewing, however, an average cup of coffee contains about twice as much caffeine as a cup of tea (100 mg versus. 50 mg). Cola nuts are also a rich source of caffeine, and the caffeine content of cola beverages typically runs 40–50 mg per 12-oz container. Many commercial analgesic (pain-killing) medicines also contain caffeine to counteract the drowsiness often produced by the active ingredients in such preparations. The most important commercial source of caffeine is as a by-product in the production of decaffeinated coffee.

3. Assume that the total daily coffee consumption in the United States averages about 1 cup per person. What is the total daily caffeine intake in the country? (Take the population as 225 million adults and the amount of caffeine per cup at 100 mg. In fact, cola drinks and medications provide additional sources of caffeine intake.)

Caffeine is easily isolated and separated from other substances present in the tea leaf or coffee bean by virtue of the fact that it is soluble in hot water, can be extracted from cold water by chlorinated hydrocarbons such as methylene chloride, and sublimes readily. In the first step caffeine is extracted from the plant source with boiling water. This separates it from the cellulose, plant proteins, fats, and other insoluble high-molecular-weight substances present in the leaf or bean. Soluble sugars, peptides, plant pigments, acids, and tannins (pentadigalloylglucose derivatives) also dissolve in the hot water. The addition of sodium sulfate increases the ionic strength, which helps prevent emulsion formation during the next step. The aqueous layer is then extracted with methylene chloride to separate the caffeine while the sugars, peptides, and polar pigments remain in the aqueous phase. Evaporation of the methylene chloride then affords crude caffeine, which is conveniently purified by sublimation. Because pure caffeine is reported to melt at 238°C and to sublime at a lower temperature, the isolated material can be characterized by converting it to a salicylate salt, mp 135–137°C. A separation and purification scheme is outlined at the end of the experiment for reference.

A. Isolation of Caffeine from Tea

1. If there are labels attached to the bags, do not tear them off. Otherwise holes can form that allow tea leaves to escape.

Place 5 tea bags[1] (about 10 g) and 200 mL of water into a 400-mL beaker and boil the mixture gently with a burner or hot plate for 15–20 min to produce a concentrated extract. *Be careful not to break the teabags.* Cool the beaker in ice until it is at or slightly below room temperature and squeeze as much solution as possible from the tea bags as you remove them by hand. (The volume of liquid should be 100–150 mL.) Dissolve about 10 g of anhydrous sodium sulfate[2] in the cold solution, and then add about 3 g (or 15 mL in a dry graduated cylinder) of Celite[TM] (filter aid).

2. The anhydrous material is much easier to weigh and handle than the decahydrate.

Fit a 5.5 cm dia. (or 4.5 cm dia.) Büchner funnel into a 125-mL filter flask connected through a trap to an aspirator or diaphragm pump (Figure E7.1), add filter paper, wet it, and turn on the vacuum for a short time to ensure that the paper is snugly pressed against the porcelain plate of the funnel. Prepare a pad of Celite[TM] by forming a slurry of about 2 g (or 10 mL in a *dry* graduated cylinder) of Celite[TM] with 10 mL of water in a small beaker. Turn on the pump or aspirator again, swirl the slurry, and *quickly* add it to the funnel. Allow it to form a uniform pad at least several mm in thickness.[3] Discard the water that has accumulated in the filter flask.

3. If the thickness of the pad is highly irregular, do it again before you proceed.

With the aspirator or diaphragm pump on, carefully add the slurry of tea extract and Celite[TM] to the Celite[TM] pad already in the funnel. When this vacuum filtration is complete, let the pad suck dry for a minute.

NOTE

The filtrate will not be clear, but will have enough particulate matter removed to minimize emulsion problems in the next step of the isolation.

4. What purpose does a filter aid such as Celite[TM] serve?

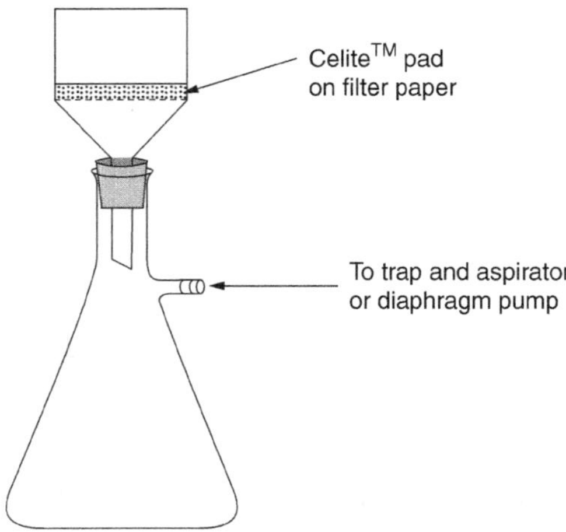

Figure E7.1 Suction filtration apparatus with Büchner funnel

Transfer the extract to a 250-mL or 500-mL separatory funnel,[4] add 25 mL of methylene chloride, and gently shake the mixture for about 2 min to ensure good equilibration between the phases. Allow the funnel to stand for 10 min. Drain the lower layer (*which will not be fully separated*) into a 125-mL Erlenmeyer flask. Extract the aqueous layer with three more 25-mL portions of methylene chloride, each time shaking gently for about 2 min and then allowing the mixture to stand for 10 min before draining the lower layer. Pour the combined methylene chloride extracts into a 125-mL separatory funnel and allow a few minutes more for most of the residual aqueous layer to separate to the top. Use the essentially clear lower layer for one of the two following isolation procedures.

Procedure 1. Add a boiling chip to the combined methylene chloride extracts in a 125-mL Erlenmeyer flask and carefully evaporate all but a few milliliters of the solvent in a hood with gentle heating (sand bath, heat lamp, steam bath, or hot plate).

Procedure 2. Evaporate all but a few milliliters of the methylene chloride layer in several portions with a rotary evaporator. A 100-mL round-bottomed flask is a convenient size to use, but never fill it more than one-third full with solvent to be evaporated.

4. If there are tea leaves present, allow them to settle and carefully decant (pour off) the upper liquid into the separatory funnel.

NOTE

Although the use of chloroform results in greater recovery of caffeine from the aqueous solution, methylene chloride is a much safer solvent to handle in quantity.

WASTE DISPOSAL NOTE

If a rotary evaporator is used, methylene chloride collected in the cold trap should be placed into a labeled container for chlorinated hydrocarbon waste.

Transfer the concentrated solution with a filter pipet to a small sublimation apparatus such as that shown in Figure E7.2. With the top of the sublimator open and with a diaphragm pump or aspirator hose connected to the side arm, carefully evaporate the methylene chloride by gently heating the bottom of the sublimator with a heat lamp or microburner. Scrape any caffeine from the sides of the sublimator into the center of the bottom with a long spatula.

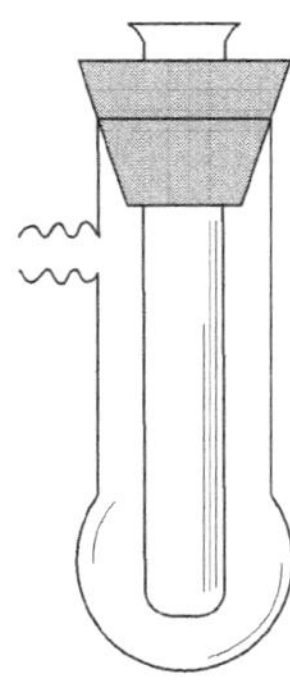

Figure E7.2 Simple sublimation apparatus

     Experiment 7   Caffeine; Isolation of an Alkaloid

Clamp the sublimator in a vertical position with the cold finger in place and the diaphragm pump aspirator connected to the side arm and turned on. Fill the cold finger with ice water, and heat the caffeine with a small flame (microburner) until it sublimes onto the cool surface. Continue until no more caffeine vapors are seen. Turn off the aspirator, allow the apparatus to cool for a few minutes, and *carefully* remove the cold finger so that very little of the sublimate falls back into the residue in the bottom of the sublimator. Scrape the sublimate onto some weighing paper and slide it into a small tared (weighed) vial. Calculate the approximate percent of caffeine recovered based on the assumed weight of the tea leaves (10 g).[5]

B. Caffeine Salicylate

Mix about 75 mg (0.543 mmol) of salicylic acid with 100 mg (0.515 mmol) of caffeine and 5 mL of methylene chloride in a 10-mL Erlenmeyer flask. Warm the flask on a steam bath until all the solid dissolves, carefully add Skelly F (petroleum ether) until the solution just becomes cloudy, heat it and add a few drops of methylene chloride to make it clear, then allow it to cool slowly. Needle crystals should form. Cool the mixture in ice and isolate the caffeine salicylate by suction filtration with a small Hirsch funnel. Allow the sample to dry and obtain a melting point (approximately 137°C).

EXERCISES

5. Write the equation for the reaction of caffeine with salicylic acid.

6. On contact with water, caffeine salicylate hydrolyzes almost completely. How does the fact that the K_b of caffeine is 7×10^{-15} in water explain this phenomenon?

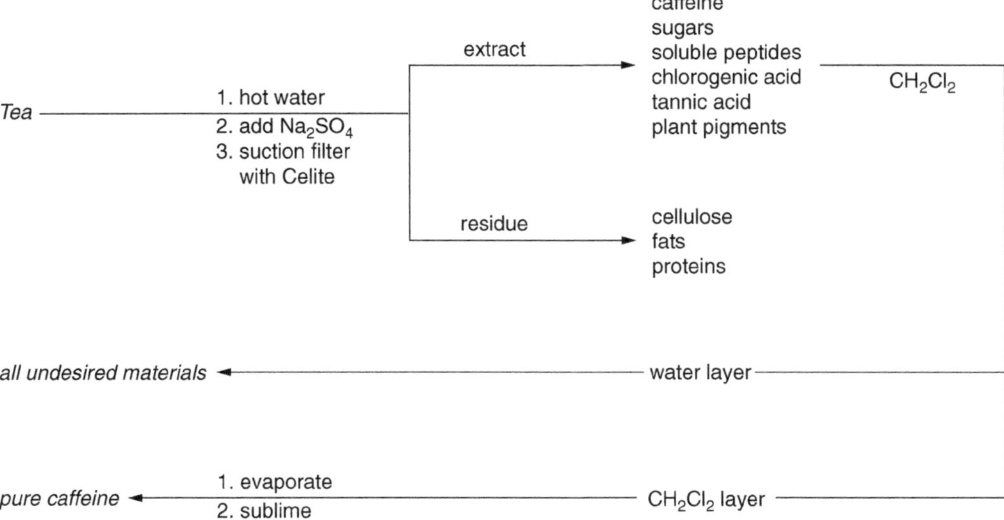

Caffeine Isolation and Purification Scheme

Experiment 7

Caffeine; Isolation of an Alkaloid

Isolation of Caffeine from Tea:

Procedure:

1. Place 5 tea bags into a 250 ml beaker. Add 150 ml water and boil for 10 minutes.
2. Cool in ice water to room temperature, remove the bags (squeeze as much solution as possible from the bags.
3. Dissolve 5g of anhydrous sodium sulfate in the cold tea solution, then add about 4 ml of celite (celite is a filter aid)
4. prepare a Buchner funnel with a filter paper.
5. prepare a pad of celite on top of the filter paper as follows:
   Mix 3 ml celite with 20 ml water in a small beaker, Turn on the vacuum pump, swirl the celite mixture with water and quickly add it to the funnel.
6. Discard the water in the filter flask and wash the flask with water.
7. With the vacuum pump on, add the tea mixture with celite (from step 3).
8. Transfer the extract to a separatory funnel and add 25 ml of methylene chloride. Gently shake the mixture for about 2 minutes, allow the funnel to stand for 10 minutes and drain the lower layer in 150 ml beaker. Extract the aqueous layer with another 15 ml of methylene chloride.
9. Wash the separatory funnel with water. Pour the methylene chloride extract into the separatory funnel and drain the lower layer in a 50 ml beaker (the lower layer contains caffeine).
10. Add a boiling chip and heat (use heating mantle) under the hood to evaporate all methylene chloride but few milliliters (about 5 ml left in the beaker).
11. Weigh a clean and dry test tube, use regular test tube (20X150 mm).
12. Transfer the concentrated caffeine solution to the test tube.
13. Stopper the test tube with a rubber stopper with one hole connected to a rubber tube through a small piece of glass tubing (see figure 1 on page 2)
14. Connect the rubber tube to the vacuum pump and turn on the pump to evaporate methylene chloride. Solid caffeine will be left behind. (reducing pressure over the caffeine solution enhances the evaporation of the remaining amount of methylene chloride)
15. Weigh the test tube and the caffeine and find the mass of caffeine recovered.

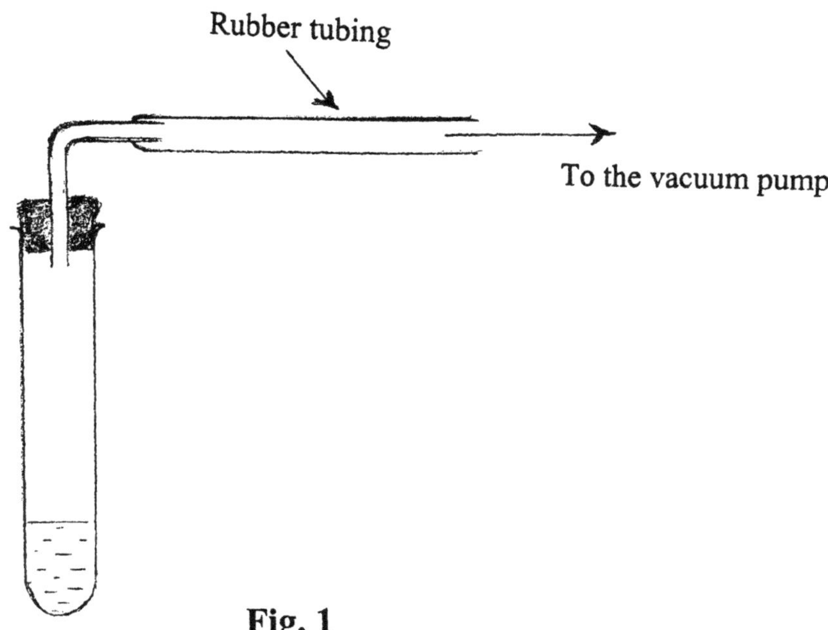

Fig. 1

Answer the following questions:

1. Calculate the approximate percent of caffeine recovered based on the assumed weight of tea leaves used of 10g.
2. Write the structural formula for caffeine.
3. Why caffeine is an alkaloid?
4. What purpose does a filter aid such as celite serve?
5. Assume that the total daily coffee consumption in the United States averages about 1 cup per person. What is the total daily caffeine intake in the country in kilograms?
   (Take the population as 250 millions adults and the amount of caffeine per cup at 100 mg).

****

Gas Chromatography

Chromatography is a method that separates a mixture into its components by distribution between two phases, one mobile and one stationary.

In a gas chromatography the mobile phase is an inert gas (Helium) and the stationary phase is usually a solid substance packed in a column.

The Gas Chromatograph : (See fig. 1)

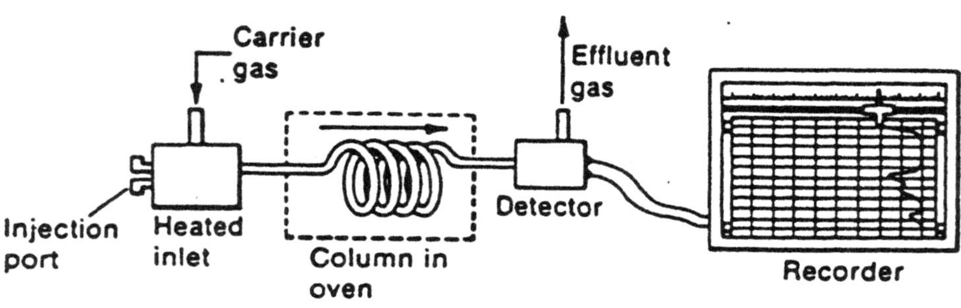

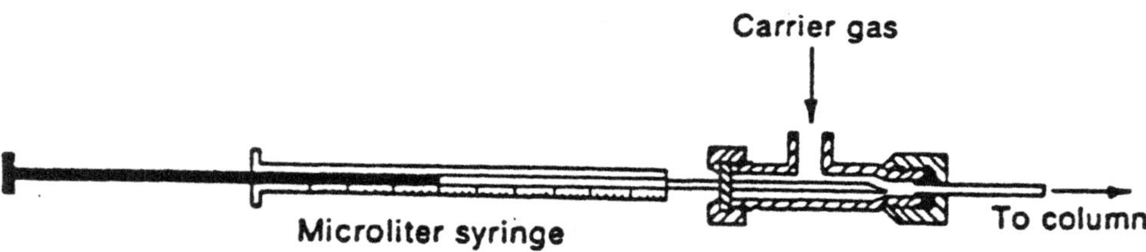

Fig. 1   Diagram of a gas-phase chromatograph

- Micro-liter syringe, the amount of sample used varies from a few tenth of μ L to 20 μL.
- Self-sealing silicone rubber septum
- Carrier gas (helium or nitrogen).
- Column, packed with solid substance.
- Detector, gives electrical signals proportional to the amount of each component in the mixture. The signals are displayed as peaks on a strip of chart. See Fig 2.

2. Retention Time: Retention time is the time a component takes to exit the G.C.
Or it is the time for a peak to appear.

<u>To calculate retention time:</u>

a) Measure the distance from the injection point (S) to the peak in cm. (See Fig 2)
b) Find the speed of the chart per minute.

Example :   If the distance   =   4 cm
            And the speed of the chart   = 2 cm / minutes .
            Find the retention time

Solution: Retention time = 4cm X 1 min / 2cm   =   2 minutes.

<u>Factors affecting retention time:</u>

Many factors play a roll on retention time, but the most important factor is
<u>polarity.</u> When a sample is injected into the G.C. it spends most of its time in the
column. Columns come in a variety of polarities and different columns are used
for different purposes.
The choice of a column for a particular separation is the most important step in
determining the success of the experiment. The basic rule about columns is
Like dissolves like, If a polar sample is injected into a polar column, the retention
time will be high but if a non- polar sample is injected into the same column its
retention time will be low. If a sample is composed of two components in which
one is polar and the other is non- polar, separation will be easy and many columns
will give good results. If a sample is a multi component mixture, column selection
is often a process of trial and error.

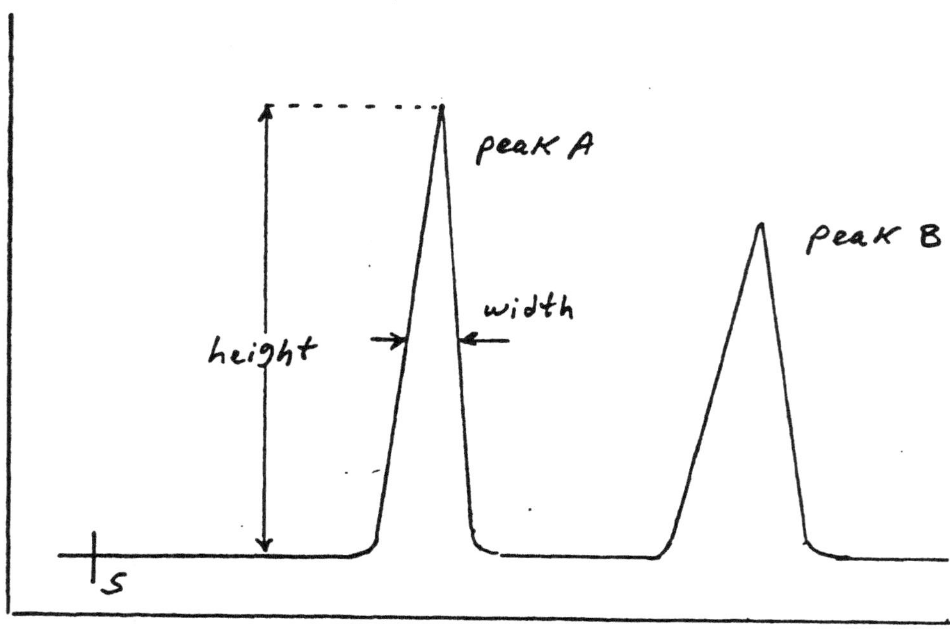

Fig 2.   Chromatogram for a mixture of two components A and B

<u>Qualitative Analysis:</u>

To find the identity of an unknown, comparison should be made between the unknown and the authentic sample on the same column. The only certain method of identification is to collect a small sample of the unknown and examine it by **infrared** (IR), **nuclear magnetic resonance (NMR)** or **mass spectral methods.**

<u>Quantitative Analysis:</u>

1. **Components of a Mixture** : Determination of the relative amounts of the compounds separated by G.C. , that results in peaks on a strip chart recorder, begins with measuring the area under each peak. (See Fig 2) The area under the peak is proportional to the concentration of the substance.
   Example: In Fig 2 :

Area under peak A   =   height X width at ½ height
Relative % of component A = area under peak A X 100 / Total area under peak A&B

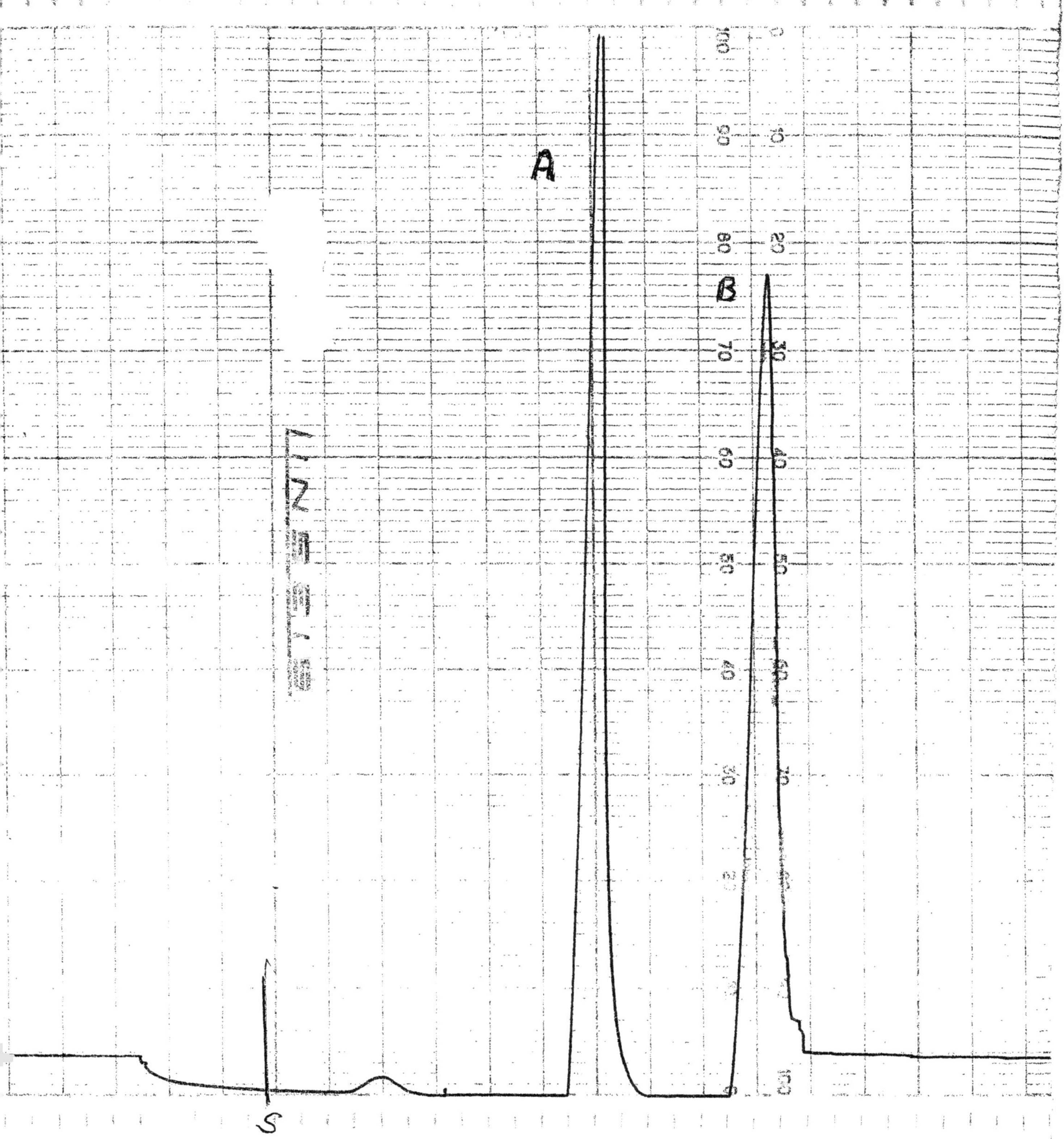

1. Calculate the relative amounts of compounds A and B as shown in the chromatogram.
2. If the speed of the chart was 2 cm/minute, calculate the retention time for the compounds A and B.

45

Fractional Distillation; Gas Chromatography

Background Reading
Review "Fractional Distillation" in Section 7.3 and "Quantitative Analysis by Gas Chromatography" near the end of Section 4.5.in the lab text

Timing
(4 h)

This experiment is most conveniently done by teams of two students each. Before the start of the experiment, each student will either be provided with or should prepare about 70 mL of a solution containing 50 mol percent each of acetone and ethyl alcohol. This requires 39 mL of acetone and 30 mL of ethyl alcohol.

A. Simple Distillation Procedure (Student I)

Set up a simple distillation apparatus such as that shown in Figure E12.1. Introduce about 70 mL of the ethyl alcohol-acetone mixture together with several boiling chips into the 150- or 200-mL distilling flask and begin the distillation slowly by heating the flask with an infrared bulb, heating mantle, or electrically heated oil bath controlled by a Variac.

NOTE

If a heat lamp is used, it should be mounted horizontally alongside the flask with no obstacle between the lamp and the flask. The flask should be shielded from behind with aluminum foil to prevent excessive heat loss. If a Variac is not available, the temperature can be adjusted by varying the distance of the lamp from the flask and by including or removing the aluminum foil shield.

Collect the first 5 mL of distillate and determine the relative amounts of ethyl alcohol and acetone by means of a gas chromatograph.

NOTE

Be certain that the sample for analysis is tightly capped during the time you are waiting to use the gas chromatograph or the composition could change as the more volatile acetone evaporates.

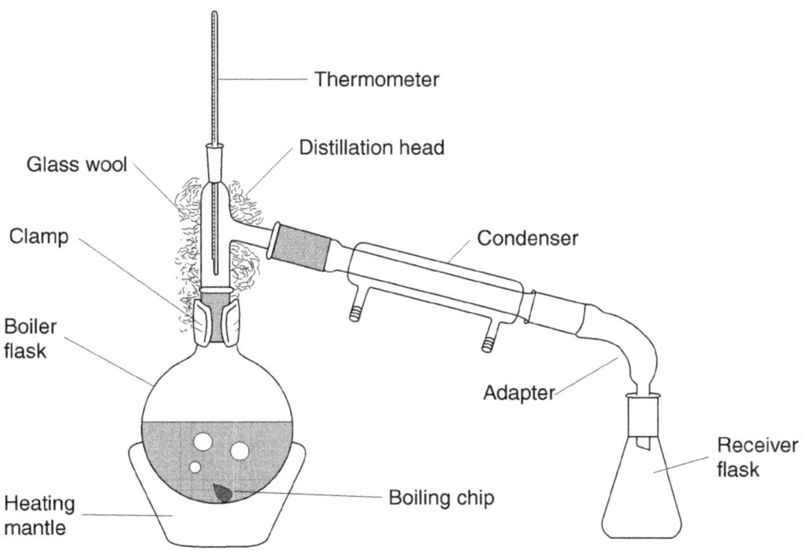

Figure E12.1   Apparatus for simple distillation

Consult your instructor about the detailed operating procedures for the chromatograph that is available to you. Either the instructor will inject your sample or you will be provided with a small syringe, which will allow you to inject several microliters of sample through a silicone-rubber septum that is part of the inlet system on the instrument. At the time of injection, mark the position of the pen on the chart paper. Be certain to record in your notebook (1) the material packed in the chromatography column (both the support and liquid stationary phase), (2) the column size (length and diameter), and (3) the approximate temperature of the column oven.

When the sample has passed through the column, two peaks should be displayed on the chart paper, together with retention times in minutes and relative areas.[1] (There may be impurities in the ethyl alcohol or acetone, which could result in additional minor peaks.)

1. If an electronic integrator is not available, simply estimate the areas by using the height of the peak times the width at half height.

EXERCISE

1. If you were not certain which peak was acetone and which was ethyl alcohol, how could you determine that?

___

Collect the remainder of the distillate in a small graduated cylinder at a rate of approximately 1 drop every few seconds. Draw a distillation curve by plotting the boiling point on the vertical axis versus milliliters of distillate (at increments of no more than several milliliters each) on the horizontal axis until a temperature of 79°C is reached.

Share data with the student doing Part **B** and perform Exercises 2 and 3.

B. Fractional Distillation Procedure (Student II)

Set up a fractional distillation column such as that shown in Figure E12.2 with an infrared bulb, heating mantle, or electrically heated oil bath controlled by a Variac as the heat source. Be certain to insulate the fractionation column and distillation head with glass wool.[2] Carry out the distillation as described for student I in Part **A** and at no more than 1 drop every 5 seconds. Obtain a gas chromatogram on the first 5-mL portion of distillate.

2. Aluminum foil should be used to hold the glass wool in place and to provide additional insulation.

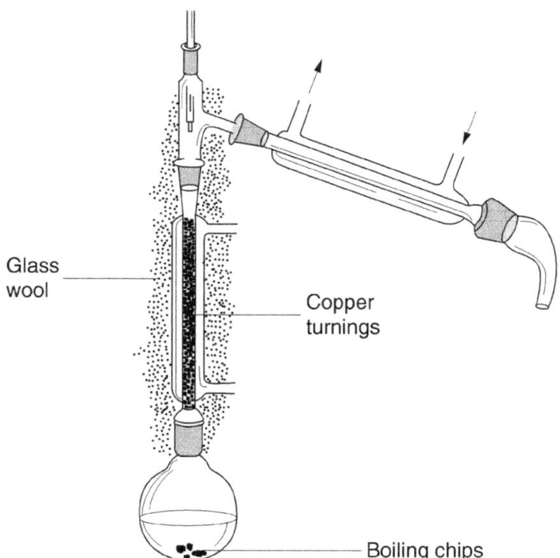

Figure E12.2 Apparatus for fractional distillation; heat source not shown

---

NOTE

In order to maximize the separation efficiency of a fractionation column, fractional distillations must be done very slowly so that equilibrium is achieved between the vapor and liquid phases throughout the column. If your initial distillate comes into the receiver at a rate greater than 1 drop/5 sec, turn down the temperature, return the distillate through the top of the fractionation column with a Pasteur pipet, and try it again.

Draw the distillation curve as described in Part **A**.

WASTE DISPOSAL NOTE FOR PARTS A AND B

Both ethanol and acetone are biodegradable and can be flushed into the drain with excess water. However, your instructor may have you collect this liquid waste for an alternate disposal method such as incineration.

C. Evaluation of Data (Both Students)

- The relative areas under the peaks for ethyl alcohol and acetone on each chromatogram will not accurately reflect the molar amounts of the two compounds. To obtain areas that accurately reflect the actual mol ratio, determine the response factors (from a chromatogram of the 1:1 molar starting mixture) as described in Section 4.5 and divide each of the areas from Parts **A** and **B** by the appropriate response factor to obtain a corrected area.

● Use the corrected data on the initial distillate from the simple distillation in Part **A** to calculate the enrichment factor, α, as described in Section 7.2. Now use the value of the enrichment factor and the mol ratio of acetone to ethyl alcohol in the initial distillate from the fractional distillation of Part **B** to evaluate the number of theoretical plates for the fractionation column (and the HETP of the column). Remember that the distillation flask corresponds to the first theoretical plate (the 1 in the exponent of the equation that follows). The appropriate equation simplifies to:

$$\alpha^{n+1} = \frac{N_{acetone}}{N_{ethanol}} = \text{mol ratio of acetone to ethanol in initial distillate}$$

Take the logarithm of both sides and solve for n.

$$n = \frac{\log(N_{acetone} / N_{ethanol})}{\log \alpha} - 1$$

Share your data with the student doing Part **A**.

2. The simple distillation is taken to be 1 theoretical plate by definition. How many theoretical plates did you find for the fractionation column? Your value should be greater than 1, but if it wasn't, how could you explain that? (*Hint: Consider the rate of distillation.*)

3. Compare the distillation curves from Part **A** and Part **B**. What are your conclusions?

Properties of Hydrocarbons

Hydrocarbons are those organic compounds composed only of carbon and hydrogen. There are *three main categories of hydrocarbons:* saturated, unsaturated, and aromatic. Saturated hydrocarbons have only carbon–carbon single bonds, whereas unsaturated hydrocarbons have carbon–carbon double or triple bonds. Aromatic hydrocarbons are cyclic compounds whose chemical properties are related to benzene.

Saturated hydrocarbons (alkanes and cycloalkanes) are relatively inert and do not react with common laboratory reagents. Unsaturated hydrocarbons (alkenes and alkynes), however, readily undergo addition reactions and oxidation reactions. Benzene and other aromatic compounds do not undergo addition reactions but are characterized by substitution reactions in which another atom or group of atoms replaces a ring hydrogen.

Although alkanes are relatively inert, they do undergo combustion in the presence of air if ignited. The fact that these reactions are highly exothermic and that huge quantities of alkanes are available as petroleum and natural gas has resulted in their extensive use as fuels. While the chemistry of alkanes is relatively straightforward, their economic impact can hardly be overestimated. Consequences resulting from this economic importance include oil spills, air pollution from automobiles, the greenhouse effect, and the threat of war in certain parts of the world.

Experiments

Performing the following tests will illustrate general properties of hydrocarbons as well as differences in chemical reactivity due to the type of hydrocarbon (saturated, unsaturated, or aromatic) being considered. Perform the first six tests on cyclohexane, cyclohexene, toluene, and on two unknowns. One unknown will be an alkane or cycloalkane and a second will be an alkene or alkyne. Using the information that a lack of reactivity is characteristic of saturated hydrocarbons and addition reactions are common for unsaturated hydrocarbons, decide the identity of each unknown as to type of hydrocarbon.

 **SAFETY NOTE** CONCENTRATED SULFURIC ACID OR BROMINE CAN CAUSE BURNS. IF EITHER IS SPILLED ON THE SKIN, WASH IMMEDIATELY WITH WATER. EXPOSURE TO BROMINE AND METHYLENE CHLORIDE VAPORS SHOULD BE AVOIDED; USE THESE REAGENTS IN A HOOD.

Procedure

1. **Solubility**   Place 2 mL of water in a 13 × 100-mm test tube and add 2 or 3 drops of the hydrocarbon to be tested. Shake the mixture to determine whether the hydrocarbon is soluble (a colorless second layer may be hard to see). Record your results and save the mixture for Test 2.
2. **Relative Density**   Reexamine the mixtures prepared above and decide in each case whether the hydrocarbon is more dense (sinks) or less dense than water (floats).
3. **Flammability**   Test the flammability of each hydrocarbon by placing 2 or 3 drops on an evaporating dish in the hood and igniting it with a match or microburner. Note the nature of the flame: more sooty flames are characteristic of unsaturated compounds.
4. **Addition of Bromine**   Dissolve 3 or 4 drops of the hydrocarbon in 1 mL of methylene chloride (dichloromethane) in a 13 × 100-mm test tube. Add dropwise a 2% solution of bromine dissolved in methylene chloride with shaking. The loss of bromine color is an indication of an unsaturated compound (do not confuse diminution of color due to dilution of the Br_2/CH_2Cl_2 solution with an actual loss of color).
5. **Reaction with Potassium Permanganate**   Dissolve 3 or 4 drops of the hydrocarbon in 1 mL of reagent-grade acetone and then add dropwise a 1% solution of potassium permanganate with shaking. A loss of the purple color of the permanganate solution indicates that a reaction has taken place and that the hydrocarbon is unsaturated.
6. **Sulfuric Acid**   Place 1 mL of concentrated sulfuric acid in a 13 × 100-mm test tube and then add 3 or 4 drops of the hydrocarbon one drop at a time. (Caution!) Reaction is indicated not only by the dissolution of the sample but also by changes in color, production of heat, or the formation of insoluble material.

At this point, information from Tests 4, 5, and 6 should allow you to determine which unknown is an alkane or cycloalkane and which is an alkene or alkyne. Record your conclusions about each unknown.

1. Set - up data sheet
2. Follow procedures **first** with knowns,   then   with your unknowns.
    be sure to give leters or numbers of each unknown and your decision
    as to what the unknown is.
3. Explain your reasons for the above decisions.
4. Answer questions # 1 - 7.

QUESTIONS

1. Considering your results in the solubility tests, what do you conclude about the solubility of hydrocarbons in water? Predict the solubility of gasoline and motor oil in water.
2. Are hydrocarbons less dense or more dense than water? Does this have any practical importance when oil spills occur?
3. Write equations using 1-butene to illustrate the reactions occurring in Tests 4, 5, and 6 of the experimental procedure (use your text).
4. Write equations using cyclohexene to illustrate the reactions in Tests 4, 5, and 6.
5. Write a balanced equation for the complete combustion of octane. What other product(s) are produced when incomplete combustion of octanes occurs in an automobile engine?
6. Write an equation to illustrate the reaction of toluene with concentrated nitric acid/concentrated sulfuric acid (2,4-dinitrotoluene is produced).
7. How could one distinguish octane from 1-octene by a simple chemical test; 1-octanol from 1-octene; and toluene from 1-octene?

Test	Cyclohexane	Cyclohexene	Toluene	Unknown......	Unknown.....
Solubility					
Relative Density					
Flammability					
Reaction with Bromine					
Reaction with $KMnO_4$					
Reaction with H_2SO_4					

Extraction and crystallization of Benzoic Acid and Benzil

- Follow Lab procedure as written on page 339.
- Benzil is used instead of Triphenylmethane.
- Answer questions # 1,2 and 3 (page 339 & 340)
- Answer the following questions:
  4. describe why vacuum drying would be an advantage if we used it.
  5. Why is a heat lamp (or oven with controlled temperature) is used instead of a Bunsen burner or hot plate?.
  6. State the function of HCl in part B.

In your report include balanced equations for reactions that might take place, and show your calculations.

Extraction and Crystallization; Acid–Base Properties

In the following experiment a mixture of equal weights of benzoic acid and triphenylmethane will be separated by extraction followed by purification of the triphenylmethane by recrystallization.

A. Separation

The sample consists of a binary mixture of about equal weights of triphenylmethane (neutral) and benzoic acid. Weigh about 3.5 g of the mixture to the nearest 0.1 g and dissolve it in 15 mL of methylene chloride (small amounts of insoluble impurities might be present).

EXERCISE

1. Calculate the theoretical amount (volume) of 3 *M* sodium hydroxide needed to convert the benzoic acid (half the weight of the sample) to its water-soluble salt. Check the results of your calculation with your instructor before you proceed.

Extract the solution in a small separatory funnel with two portions of 3 *M* sodium hydroxide such that the volume of each portion is about 1.5 times the theoretical amount calculated in Exercise 1. To avoid forming an emulsion, do not shake the separatory funnel vigorously, but use a gentle, swirling motion. *Remember to hold the separatory funnel with both hands (Figure E10.1) and to vent it frequently with the lower end pointed upward and away from other persons.* Allow the funnel to stand for a while before draining the lower methylene chloride layer. *Be certain to label and save both layers.*

EXERCISE

2. If you were not certain which layer was methylene chloride, suggest a simple procedure you could use to test your assumption.

Background Reading
Section 6.1, "Standard Scale Procedures" and "Utilizing Acid–Base Properties in Extractions," and Section 5.2, "Standard Scale Techniques" for crystallization in the laboratory text

Timing
(5 h)

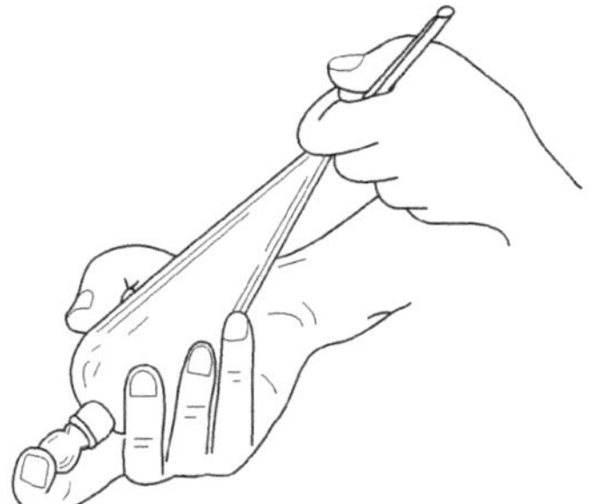

Figure E10.1 Holding the separatory funnel

Combine the aqueous extracts and carry out an extraction with two 5-mL portions of methylene chloride to ensure removal of any neutral compound suspended or dissolved in the water. The aqueous extract might be cloudy because of very small droplets of suspended methylene chloride, but these will settle eventually and will not affect the remainder of the experiment.

EXERCISE

3. Assume that concentrated hydrochloric acid is 12 *M* and calculate the volume needed to neutralize your basic extract. Check the calculation with your instructor before you proceed.

B. Isolation of Benzoic Acid

Convert the aqueous solution of sodium benzoate to benzoic acid by slowly adding a slight excess of concentrated hydrochloric acid until the solution is acidic to pH paper. Cool the suspension in an ice bath along with a second small flask or beaker containing a few mL of water. Suction filter the benzoic acid on a suitable Büchner or Hirsch funnel as shown in Figure E10.2. Rinse the beaker with a little cold water to collect as much of the benzoic acid as possible. After the acid has been sucked dry for at least 5 min, spread the sample on a piece of glazed weighing paper and dry it with gentle heating by a heat lamp from above, but take care not to melt the sample. Obtain a mp on the dry sample.

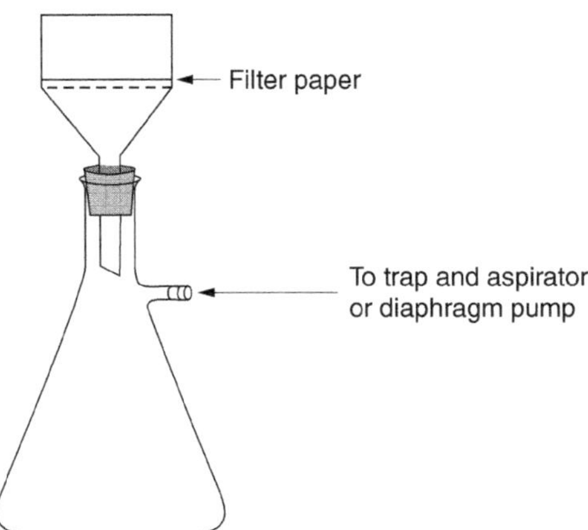

Figure E10.2 Suction filtration apparatus with Büchner funnel

As an optional experiment, recrystallize the benzoic acid from hot water by using about 15 mL of water per g of acid. Filter the crystalline product with suction, dry it, and obtain the mp.

C. Isolation and Crystallization of Triphenylmethane

Combine all the methylene chloride layers, add about 0.5 g of anhydrous sodium sulfate, and swirl the resulting suspension for a few minutes. Be certain to rinse containers from which the methylene chloride solutions are poured with a few mL of fresh solvent. Weigh a small round-bottomed flask to the nearest 0.1 g and carefully decant (pour off) the solvent into the flask. Use a few mL of methylene chloride to rinse the drying agent, and decant this solvent into the same flask. Evaporate the solvent with a rotary evaporator or as shown in Figure 2.18 and weigh it to obtain a crude yield.

To recrystallize the triphenylmethane, dissolve the crude sample in about 7.5 mL of hot hexane (or Skelly B) per g of compound, and filter the solution hot by gravity. Add a boiling stick, concentrate the solution to about one-third to one-half of the original volume, and allow the solution to cool slowly; then cool the flask in ice. Collect the crystalline triphenylmethane by suction filtration on a suitable size of Hirsch or Büchner funnel (Figure E10.2), and rinse the sample with a very small amount of cold hexane (or Skelly B). Obtain an mp. Record the percent of recovered crude and recrystallized triphenylmethane (based on half the weight of the original sample.)

EXERCISE

4. Describe how the extraction would be done with only a few hundred mg of mixture.

WASTE DISPOSAL NOTE

Hexane should be placed in a container for nonhalogenated waste solvents.

     Experiment 10   Extraction and Crystallization; Acid–Base Properties

Extraction & Crystallization
Flow Chart

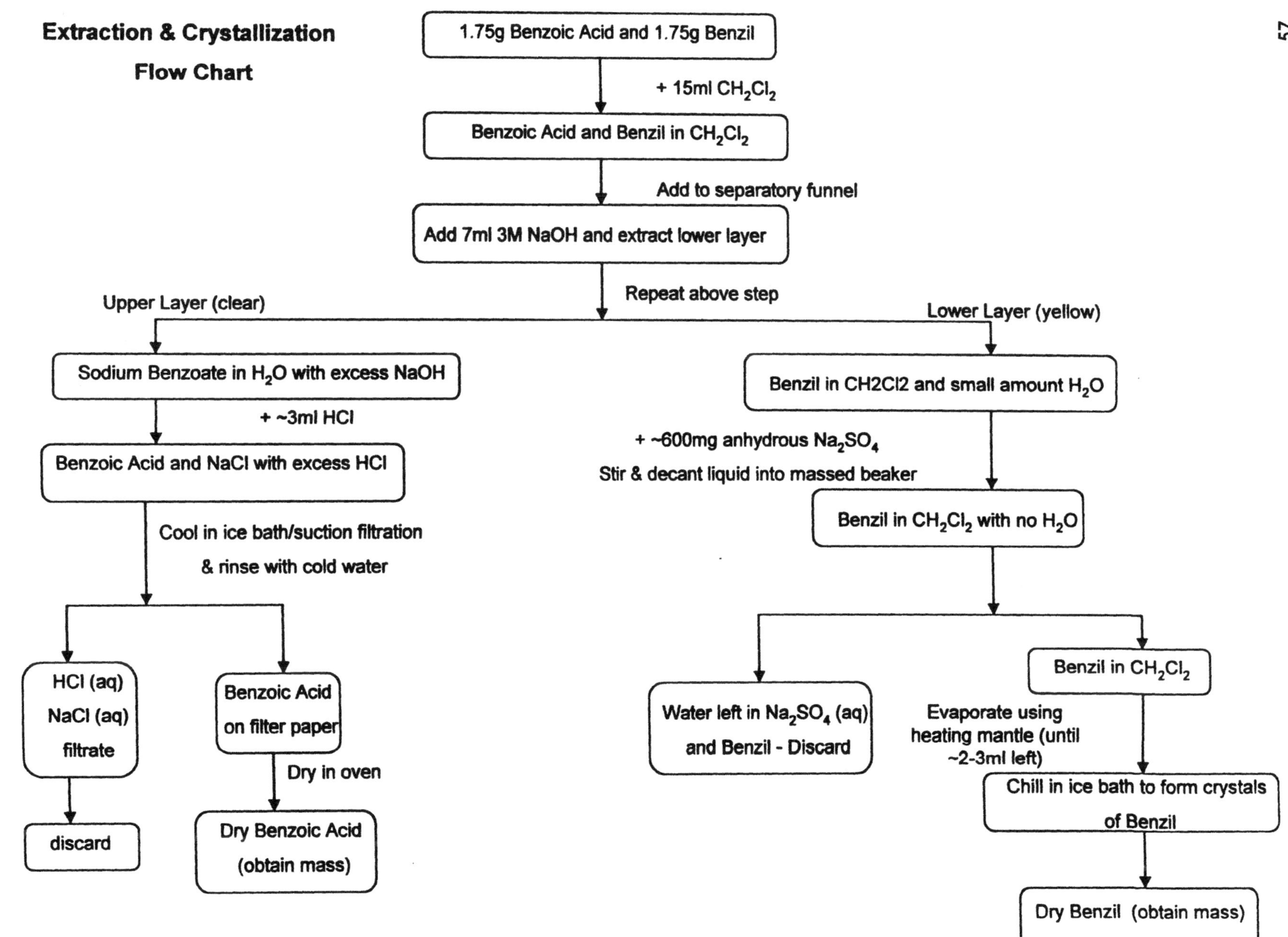

Triphenylcarbinol— Addition of a Grignard Reagent to a Ketone

Background Reading
The Grignard reaction in any organic textbook

Timing
(4-5 h)

Developed by Victor Grignard, who received the Nobel Prize for his efforts in 1912, the reaction between an alkyl (or aryl) halide and magnesium in ether, known as the **Grignard reaction**, produces an alkylmagnesium (or arylmagnesium) halide, one of the most versatile synthetic reagents available to the organic chemist. The general reaction is written

$$RX + Mg \xrightarrow{Et_2O} RMgX$$
$$\text{Grignard reagent}$$

Diethyl ether, the usual solvent for such reactions, is more than a mere solvent; it is essential to the success of the reaction. Lone pairs of electrons on the oxygen atoms of two molecules of ether apparently coordinate with the magnesium atom in the alkylmagnesium halide to form a relatively stable dietherate complex, which is frequently not shown in the equations describing the reactions of Grignard reagents.

EXERCISES

1. Evidence has been found that RMgR is also present in ether solutions prepared from an alkyl halide and magnesium. Suggest a reaction for its formation.

2. The reaction of 1,2-dibromoethane with magnesium produces a gas! What is it? Write a balanced equation for its formation.

Grignard reagents are readily formed from primary, secondary, or tertiary alkyl chlorides, bromides, and iodides, and from aryl bromides and iodides, but only with great difficulty from aryl, vinyl, and cyclopropyl chlorides. Kept out of contact with oxygen and moisture, Grignard reagents can be stored for long periods of time as ether solutions; however, it is common practice to utilize Grignard reagents immediately after their preparation.

Of crucial importance to the successful production of Grignard reagents is the absence of moisture, which can inhibit the union of the organic halide and the magnesium by forming an impervious layer of oxide and hydroxide on the magnesium surface. The reaction flask to be used is often dried in an oven or heated with a burner to drive moisture from the pores of the glass. The alkyl halide, which may contain traces of hydrogen halide as well as moisture, can be conveniently purified and dried by passage through a short column containing activated aluminum oxide (alumina). The ether can be purchased dry or can be dried by allowing it to stand over anhydrous calcium chloride for a few hours, but if the Grignard reaction is of a type that is particularly sensitive to moisture, the ether should be distilled from lithium aluminum hydride and stored over activated molecular sieves.

Under the most ideal circumstances, Grignard reactions should be carried out in an atmosphere of dry, inert gas, such as argon or nitrogen, so as to exclude the moisture or oxygen of the air from the reaction vessel. However, many reactions can be done successfully by simply attaching a drying tube containing anhydrous calcium chloride or calcium sulfate (Drierite^{TM}) to the top of the condenser. Because of the ether vapor in the flask, most of the oxygen that might otherwise be a problem is excluded and there is no need for a blanket of inert gas.

EXERCISE

3. Once formed, Grignard reagents react with water and with oxygen followed by water to produce hydrocarbons and alcohols, respectively. Write balanced equations for these two reactions.

In the following experiment, bromobenzene is converted to phenylmagnesium bromide, which then reacts with benzophenone to give triphenylcarbinol.

$$PhBr + Mg \longrightarrow PhMgBr$$

$$PhMgBr + PhCOPh \longrightarrow Ph_3COMgBr$$

$$Ph_3COMgBr + HOH \longrightarrow Ph_3COH + MgBrOH$$

A. Preparation of Apparatus

Warm a dry 10-mL pear-shaped flask, magnetic stir bar, and dry condenser in the oven (in a labeled beaker) or with a heat lamp for about 15 min. Prepare a drying tube from an eye dropper containing a small amount of anhydrous $CaCl_2$ between two cotton plugs. Except for the plastic syringe, assemble the warm apparatus as shown in Figure E22.1. Try the magnetic stirrer to make certain the stirring bar is small enough that it turns freely. Do not use grease on the joints. You will need a dry filter pipet calibrated by making several marks at 4-mm intervals (about 0.1 mL between marks) to handle dry diethyl ether. (This works better than a Pasteur pipet for transfer of liquids with a high vapor pressure.)

B. Phenylmagnesium Bromide

Lightly sand some magnesium ribbon, cut off about 3 cm, and weigh it to the nearest mg. You need a minimum of 23 mg (0.95 mmol), but avoid an amount in excess of 27 mg. Always handle the ribbon by the edge or with forceps. Disconnect the flask, clamp it, and insert a small dry funnel. With a calibrated filter pipet add about 800 µL of commercial dry diethyl ether (from a freshly opened container). Then cut the Mg with scissors into 2- to 3-mm-long pieces so that they fall through the funnel into the flask.

Gently turn on the water in the condenser, begin magnetic stirring, and heat the ether to reflux. With an automatic pipet, deliver exactly 94 µL (141 mg, 0.90 mmol) of bromobenzene (previously purified by passage through a short column of activated

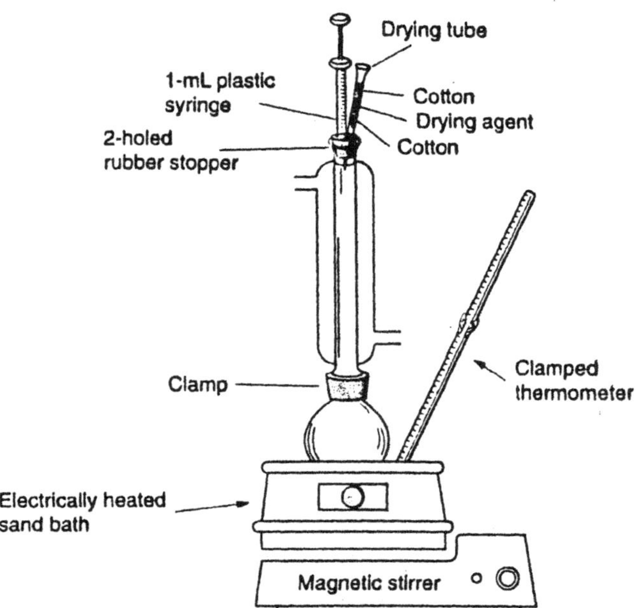

Figure E22.1 Apparatus for Grignard reaction (stirring bar not visible)

Al_2O_3 in the hood) to a small *dry* vial, and then add about 500 μL of dry ether (a freshly opened containers is preferred) with the calibrated filter pipet. Draw this solution carefully (*slowly*) into the 1-mL plastic syringe and place the needle through the second hole in the rubber stopper. Add 5–8 drops of the bromobenzene solution with the syringe dropping funnel. Continue to apply gentle heat to help the reaction start (as evidenced by a cloudy appearance; you may have to turn down the heat after the reaction starts because it is exothermic.) A few more drops of the bromobenzene may be needed. When the reaction has started, add the remainder of the bromobenzene solution slowly (15–20 min). Continue to heat the reaction mixture gently for 15 min with stirring. Only a small amount of the original Mg should remain. Remove the sand bath, but leave the apparatus on the magnetic stirrer. Allow the Grignard solution to cool.

NOTE

Although it is unlikely they will be needed, other techniques that have been used to start Grignard reactions include the following: (1) Add a very small crystal of iodine to the ether (light red-brown color); (2) crush some of the magnesium turnings (under the ether) with a stirring rod against the inside of the flask.

EXERCISE

4. Iodine, which reacts with magnesium to form magnesium iodide, has sometimes been used to treat the magnesium first. What is the likely reason why the iodine helps to start the Grignard reaction? Why would the use of 1,2-dibromoethane or crushing magnesium under the surface of the ether be effective?

C. Triphenylcarbinol

Tare a small *dry* vial to the nearest mg, and weigh 124 mg (0.68 mmol; based on an assumed yield of Grignard reagent of at least 75%) of benzophenone (finely powdered) into it. Add about 500 μL of dry ether (with the calibrated filter pipet) cap the vial, and carefully swirl the solution until the benzophenone has dissolved. Draw the solution into the plastic syringe. Magnetically stir the Grignard solution (without heating) and add the benzophenone slowly (30 sec). The magnesium salt of the triphenylcarbinol will precipitate, and the reaction mixture may solidify

D. Product Isolation

With the reaction mixture at about room temperature, slowly add $3M$ HCl (stir with a spatula) until two layers are formed and the lower aqueous layer is distinctly acid to pH paper. Allow any unused Mg to dissolve. With a Pasteur pipet carefully transfer the lower aqueous layer to a 1-dram vial and extract the layer with three 0.5-mL portions of diethyl ether (*not the special dry ether*).

After each extraction, transfer the lower layer with a Pasteur pipet to a second vial and combine the ether layer with the original ether layer by means of a filter pipet. The combined ether layer is then washed with 0.5 mL of water, which is removed and discarded. The ether solution should then be dried with about 300 mg of anhydrous Na_2SO_4 for 10 min and then rinsed carefully into a clean, tared 1-dram vial in small amounts followed by careful evaporation of the ether (boiling stick) with a sand bath to give a solid residue of the product mixed with biphenyl, a reaction by-product.

Cold ligroin (bp 30–60°C) or petroleum ether (0.5 mL) is then added to the crude product to selectively dissolve the biphenyl. Scrape and stir the suspension for a few minutes followed by removal of the solvent with a filter pipet. Rinse the crude triphenylcarbinol with an additional 0.5 mL of the solvent, and allow it to dry before determining the crude yield or taking a mp.

NOTE

The pink color commonly observed is due to very small amounts of a radical anion formed by transfer of an electron to benzophenone. Such radical anions are known as ketyls.

NOTE

Always place the vials in a small beaker so they cannot be tipped over accidentally.

Introduction to Infrared Spectroscopy

A. Operation of the IR Spectrophotometer

Background Reading
Sections 8.1–8.2 and Chapter 10 in the laboratory text, especially Sections 10.1–10.2, "Sampling Techniques," and "Use of Teflon Tape"

Timing
(4–5 h)

Small groups of students will be given instruction on how to take an infrared spectrum on a solid or liquid sample by the use of Teflon tape.

B. Identification of Functional Groups

You will receive a small sample of a solid or liquid unknown and its molecular formula. Obtain an infrared spectrum and then list on the chart each of the following things.

1. Your name
2. The code number and molecular formula of the sample
3. The IHD (Index of hydrogen deficiency)
4. Make reasonable assignments (such as aliphatic C—H stretch, O—H stretch, C—H bend, etc.) for all significant absorptions in the region 1350 cm^{-1}–3500 cm^{-1}.
5. Copy each of the following functional groups into a box on your chart and indicate whether that group is probably or definitely present (P), probably or definitely absent (A), or that it is not possible to tell (?).

RNH_2	C=C (aliphatic)	RCHO	RCO_2R	RCN
R_2NH	C=C (aromatic)	RCOR	$RCONH_2$	ROH
R_3N	C—CH	RCOOH	RCOCl	

C. Infrared Problem Set

For this part of the laboratory you are encouraged to work in small groups. Copies of 20 spectra will be made available with the correct compound included among the group of structures below each spectrum. (*Do not write on the problem sets; they must be returned at the end of the period.*) For each problem, indicate the correct structure in your notebook and briefly explain your choice in terms of the spectrum. You may use tabulated infrared spectral data in Tables 10.1–10.4 and the charts (Figure 10.30) at the end of Chapter 10.

Spectroscopy

Spectroscopic data provides chemists with information that reveal the structure of organic compounds.

The ability to analyze spectroscopic data is a skill that is gained by familiarity. The more spectra one looks at, the better one will be equipped for obtaining information from a spectrum of an unknown.

Infrared Spectroscopy:

Every compound has a unique infrared spectrum. It is possible to identify a compound by comparing its spectrum to that of known substances. The infrared spectrum also allows one to tell what functional groups are present in the molecule.

The Infrared Spectrum :

Absorption of infrared radiation can excite molecular vibrations and rotations from the ground state to an excited state. The individual bonds are bent or stretched.

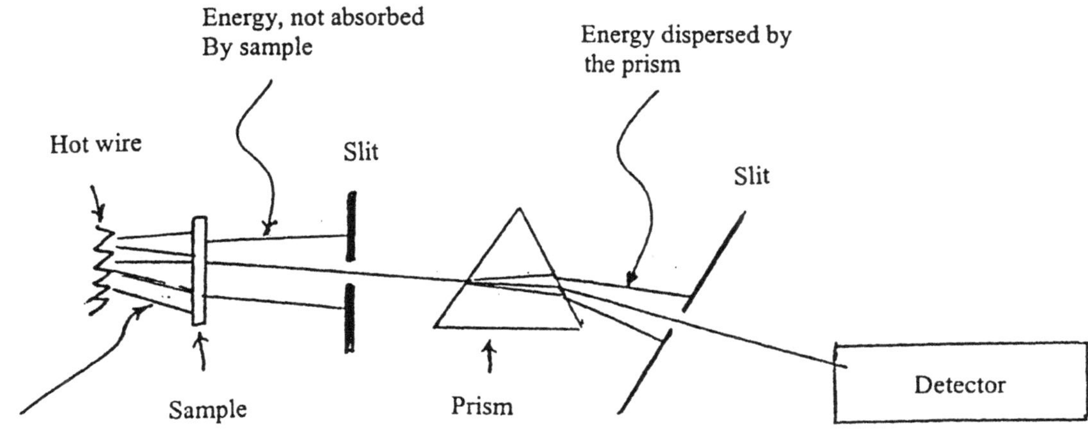

The major parts of an infrared spectrometer

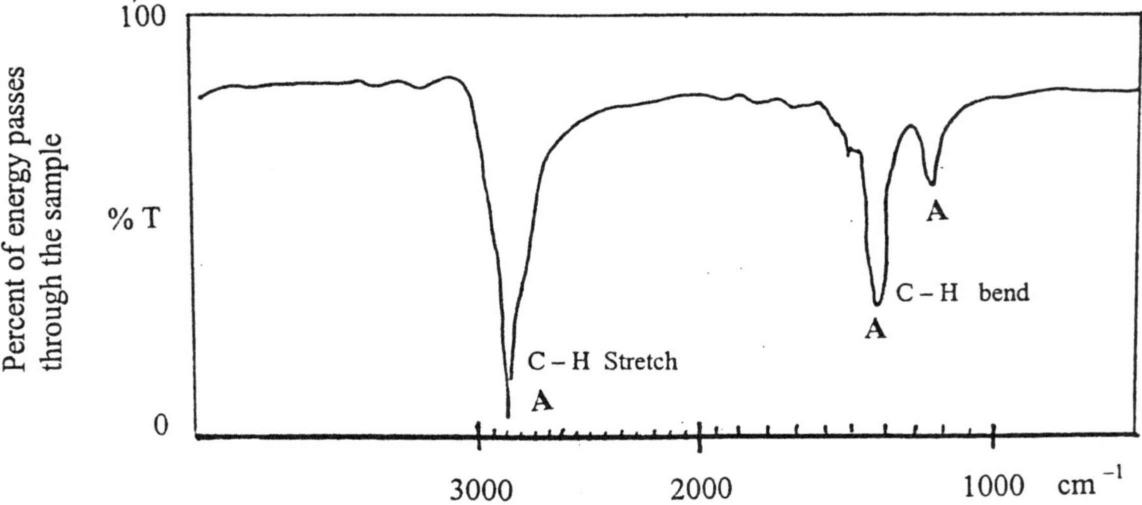

Infrared spectrum of an alkane

A: Lowest point correspond to absorption of the energy.
C – H stretch: This band is due to the absorption of energy for the excitation of carbon – hydrogen bond stretching.
C – H bend : This band result from the absorption of energy for the excitation of carbon – hydrogen bond bending.

Measurement :

The absorption of infrared radiation by the sample can be measured as the energy of the infrared beam is changed. **A plot of absorption vs wave number is called infrared spectrum.**

The energy that corresponds to the absorption of infrared radiation can be expressed as:

$$\Delta E = h\nu$$

where:

ΔE = The difference in energy between two vibrational energy levels.
h = Planck's constant ($= 6.63 \times 10^{-34}$ Js).
ν = Frequency.

The designation of an absorption peak in an infrared spectrum is described in terms of the wave number (ν) in cm $^{-1}$.

$$\nu = 1 / \lambda \qquad (\lambda = \text{wavelength}) \quad \dots\dots\dots\dots\dots\dots 1$$

Since $\qquad \lambda \nu = c \qquad (c = \text{velocity of light} = 3 \times 10^{8} \text{ m/s})$

Therefore $\quad \nu = c / \lambda \qquad \dots\dots\dots\dots\dots\dots\dots\dots\dots\dots\dots\dots\dots 2$

and since $\quad \Delta E = h\nu \qquad \dots\dots\dots\dots\dots\dots\dots\dots\dots\dots\dots\dots\dots 3$

From equation 2 and 3

$$\Delta E = h c / \lambda$$

$$\Delta E = h c \, 1/ \lambda$$

$$\Delta E = h c \nu \qquad \dots\dots\dots\dots\dots\dots\dots\dots\dots\dots\dots\dots 4$$

Equation 4 shows the proportionality between ΔE & ν

Functional Groups:

The typical organic compound has many absorption bands in the IR because of the large number of possible vibrations and rotations. However, many functional groups have absorption frequencies which can be used for identification because they tend to be at the same place in most compounds.

See the attached figures

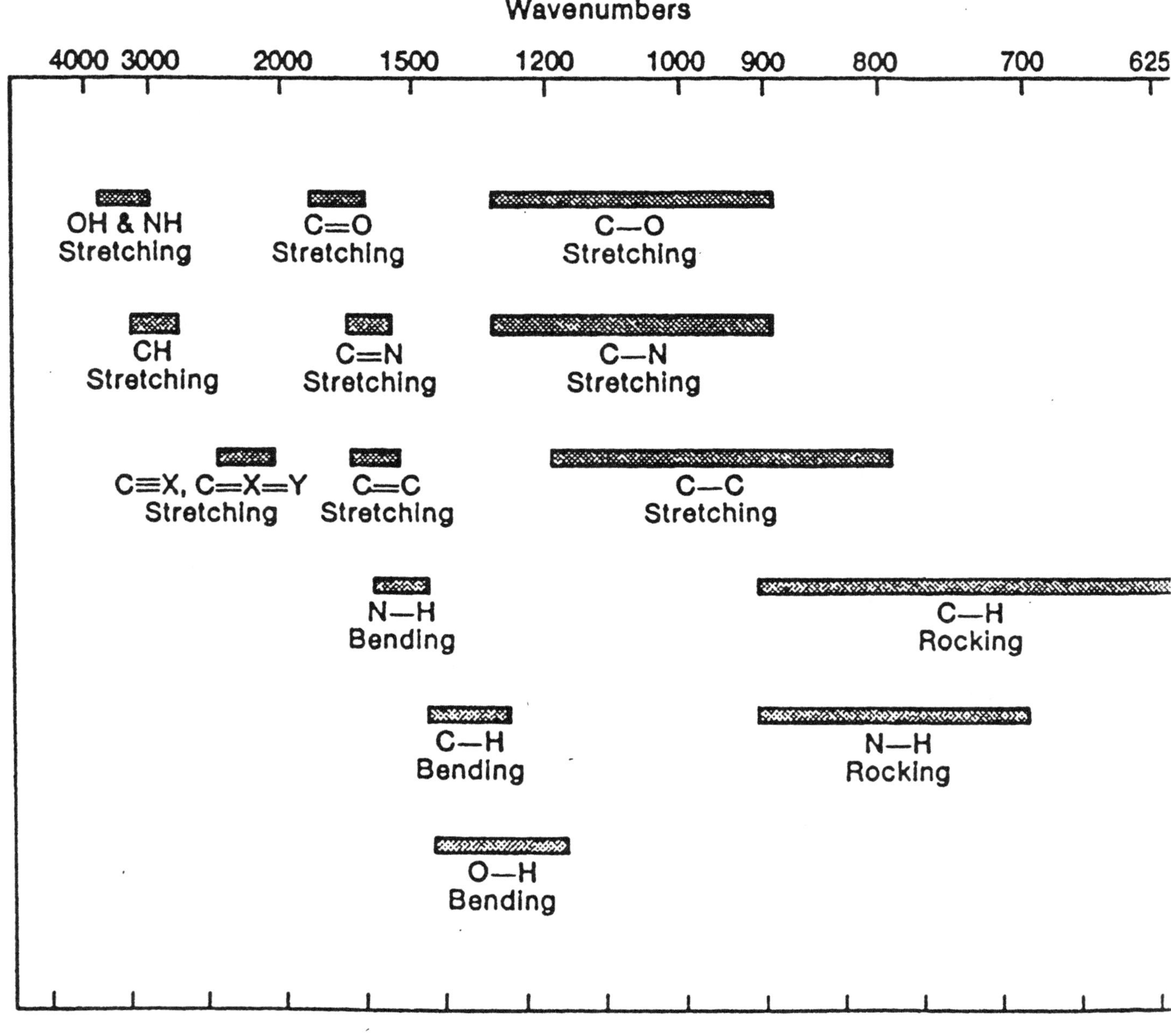

Wavenumbers
4000 3000 2000 1500 1200 1000 900 800 700 625
OH & NH Stretching
CH Stretching
C≡X, C=X=Y Stretching
C=O Stretching
C=N Stretching
C=C Stretching
N—H Bending
C—H Bending
O—H Bending
C—O Stretching
C—N Stretching
C—C Stretching
C—H Rocking
N—H Rocking
Microns

Selected Characteristic Infrared Absorptions

Functional Group or Type of Bond	Wavenumber (cm⁻¹) and Intensity[a]	Remarks
—C—H Alkanes	2980–2850 (s) 1470–1450 (m) 1465–1440 & 1390–1360 (m)	C—H stretch Methylene C—H bend Methyl C—H bends; a doublet at 1390–1360 implies a *gem*-dimethyl group
=C—H Alkenes	1680–1620 (v) 3100–3000 (m) 1000–980 & 920–900 (s) 900–870 (s) 980–950 (s)	C=C stretch; becomes weaker as degree of alkyl substitution increases; conjugated dienes often show two bands, 1650 & 1600 C—H stretch similar to aromatic C—H stretch $RCH=CH_2$; C—H out-of-plane bend $R_2C=CH_2$; C—H out-of-plane bend *trans*-RCH=CHR; C—H out-of-plane bend; no absorption in this region for *cis*-isomer
≡C—H Alkynes	2260–2100 (v) 3300–3200 (s)	C≡C stretch; strongest for terminal acetylenes C—H stretch; sharp band
Benzene ring	1600 & 1500 (v) 1580 (m) 3100–3000 (m)	Phenyl-ring vibrations; 1500 peak usually stronger than 1600 peak when ring is conjugated with unsaturated group or atom with pair of electrons C—H stretch; frequently several bands; similar to alkene C—H stretch
C=O Aldehydes, ketones, & acid derivatives	1850–1780 & 1770–1710 (m-s) 1810–1790 (s) 1765–1720 (s) 1725–1695 (s) 2820 & 2720 (v) 1725–1705 (s) 1710–1680 (s) 3300–2400 (s, b) 1680–1640 (s) 1650–1610 (m) 1560–1530 (s) 3500–3200 (m)	Anhydride C=O stretch; two bands; variable relative intensity Acid chloride C=O stretch Ester C=O stretch Aldehyde C=O stretch Aldehyde C—H stretch with C—H bend overtone (called a Fermi doublet, see text) Ketone C=O stretch Carboxylic C=O stretch (dimer) Hydrogen-bonded O—H and combination bands characteristic for COOH Amide C=O stretch (Amide I band) NH_2 deformation in 1° amides (Amide II band) N—H bending in 2° amides (Amide II band) N—H stretching in primary and secondary amides; appears as doublet in primary amides

Values given for C=O stretching frequency may deviate by ±10 cm⁻¹ in various simple structures. Hydrogen bonding to oxygen may cause a shift as large as −50 cm⁻¹. Conjugation causes a shift of −20 to −40 cm⁻¹.

Functional Group or Type of Bond	Wavenumber (cm⁻¹) and Intensity[a]	Remarks
O—H, N—H Alcohols, phenols & amines	ca. 3650 (v) 3400–3200 (s, b) 3500–3200 (m, b) 1650–1590 (s, b)	O—H stretch; sharp band from unassociated group; now always seen O—H stretch; hydrogen bonded group N—H stretch; appears as a doublet for NH_2 NH_2 deformation in 1° amines

[a] s = strong; m = medium, w = weak; v = variable; b = broad

Infrared Spectroscopy

Study the spectra of the following compounds:

a) Benzyl alcohol
b) 1- propanol
c) 3- pentanone

Note the location of the absorption bands due to the following groups:

OH stretch, C – O, C – H, C = O and C = C

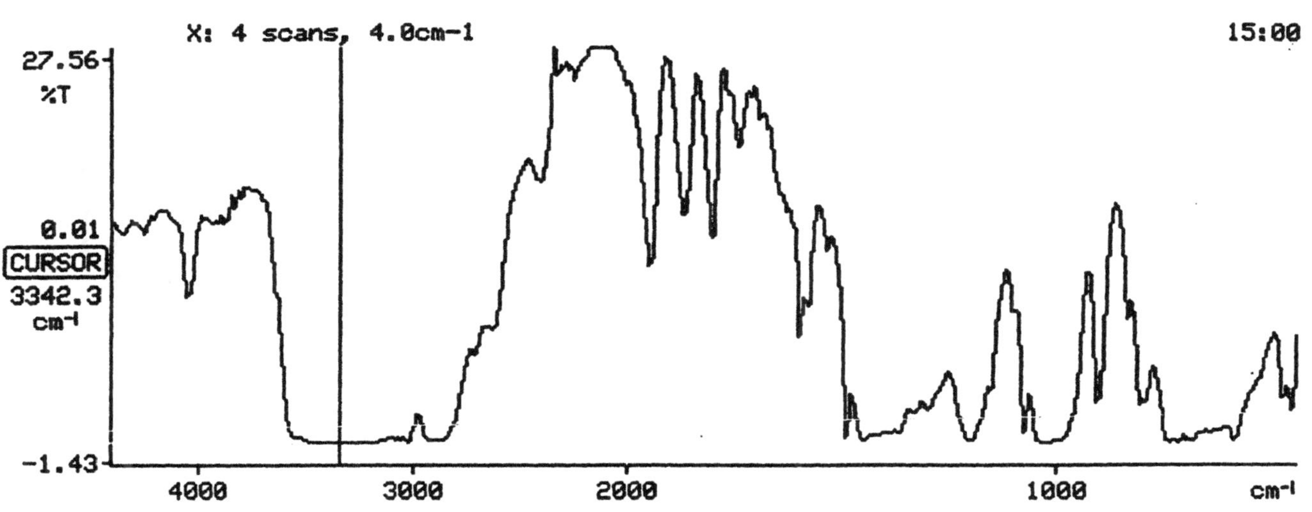

1 – Benzyl alcohol
CH₂OH

1- propanol

$-C-C-C-OH$

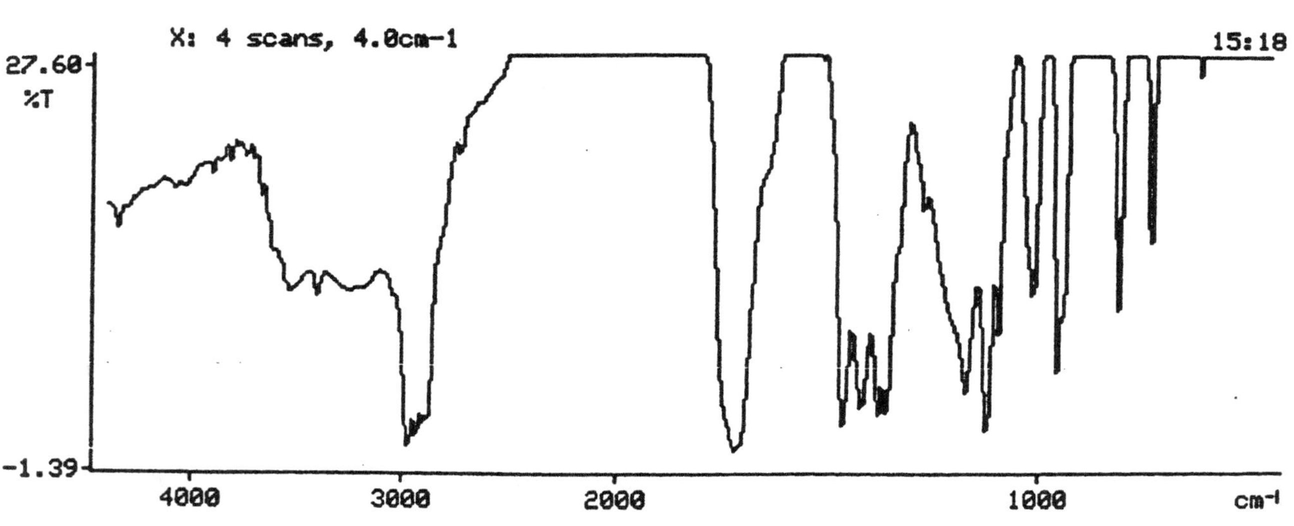

3 - pentanone

$-C-C-C-C-C-$

69

Nuclear Magnetic Resonance
<u>NMR</u>

A proton nuclear magnetic resonance spectrum allows you to determine the number and types of protons in a molecule. In addition it often allows you to tell how many protons are adjacent to a given proton. Often it is possible to determine the structure of a molecule by inspection of the NMR spectrum.

The absorption of energy, by a particular atomic nuclei, result in a change in the distribution of these nuclei among the various possible energy levels available to them.

Every nucleus has a spin quantum number which depends on the total number of protons and neutrons and may have the value 0, $1/2$, 1, $3/2$, 2, ...

The C^{12} and O^{16} found in great number of organic molecules have spins of zero and do not give rise to an NMR signal, therefore they do not interfere with or complicate the signals from other nuclei such as the proton.

A spinning charged body (a nucleus) has a magnetic field. If an external magnetic field is applied there will be two possible energy levels for this nucleus:

	Excitation	●————	high energy level
	→		
	←		
●————	Relaxation	————	low energy level

The establishment of an equilibrium between excitation and relaxation, known as the **resonance condition**, occurs in an NMR spectrum when the externally applied magnetic field and frequency (in the radio-frequency region) are properly adjusted. Under this circumstances, the slight excess of nuclei that occupy the lower of the two energy levels can be detected, amplified, and displayed as a signal on a chart.

The exact position of the resonance signal on the chart is referred as the **chemical shift** and is usually measured to a standard peak displayed on the NMR chart. The standard used for routine proton spectra is **Tetramethylsilane (TMS)**, $(CH_3)_4$ Si. The TMS produce a sharp signal peak assigned the value $\delta = 0$.

Range of δ	Type of Hydrogen	Range of δ	Type of Hydrogen
0–1.7	Aliphatic C—H	4.5–7.0	$C{=}CH$
1.7–3.0	CH_2—Ar, CH_2—NR_2,	7.0–8.0	ArH
	CH_2—CO, CH_2—S, $C{\equiv}C$	9.5–10.0	RCHO
3.0–4.5	CH_2—F, CH_2—Cl, CH_2—Br,	10.0–12.0	RCO_2H
	CH_2—I, CH_2—OH, CH_2—OR		

(table 1) Approximate chemical shift values for various types of hydrogens

The magnetic field at the nucleus of CH_3 proton is not the same at OH proton, because **the field at the nucleus depends on the shielding effect of the surrounding electrons.** The nucleus is shielded from the external magnetic field by the surrounding electrons. Therefore :

$$\text{Field at the nucleus} \;=\; \text{Applied field - shielding effect}$$

If the radiation frequency were kept constant and the magnetic field increased, we get **TWO** signals from methanol because the shielding effect of electrons associated with CH_3 is different from that of OH, so you need different external fields to get the same field at the nucleus.

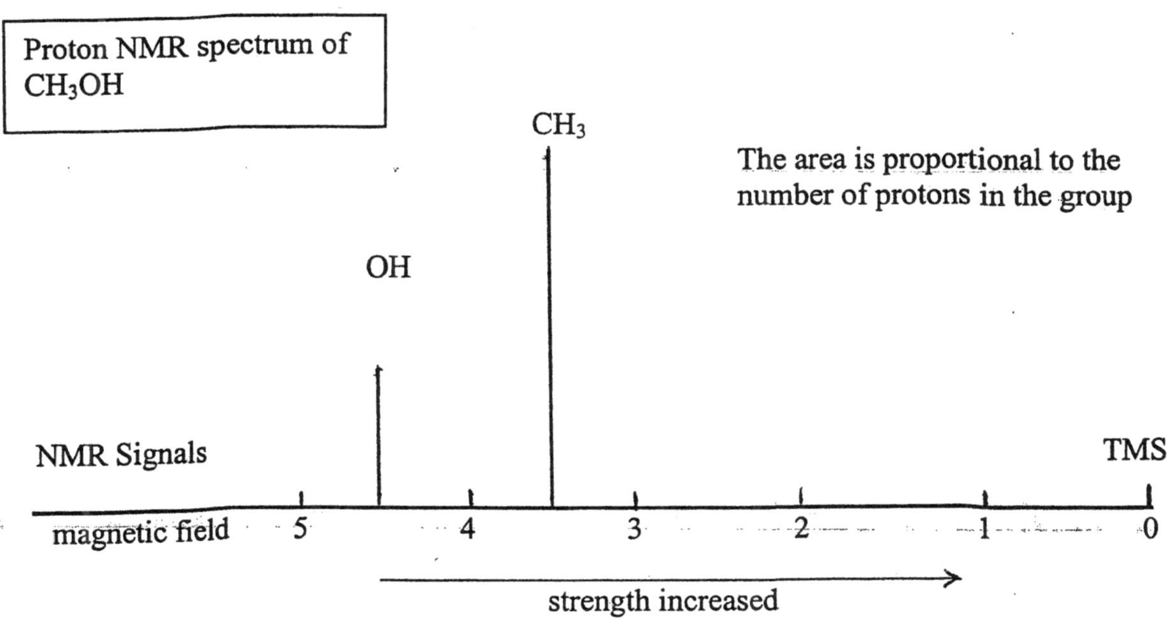

The position of an NMR signal for a proton is measured relative to the compound: Tetra methyl silane (TMS).

Chemical shift table

Group	Example	Chemical shift (δ)
CH_3 group on the end of alkyl chain	$R\text{-}CH_2\text{-}CH_2\text{-}CH_3$	0.9 - 1.5
CH_3 group adjacent to Br	$CH_3\text{-}Br$	2 - 3
CH_3 group with an adjacent carbonyl group	$R - \overset{\displaystyle O}{\overset{\|}{C}} - CH_3$	2 – 2.7
CH_3 group adjacent to ether	$R - CH_2 - O - CH_3$	3.1 – 4
A proton on carbon - carbon double bond	$R - CH{=}CH - R$	4.5 - 6
A proton on a benzene ring	(benzene ring)	7 - 8

Spin – Spin Splitting

Protons can influence one another magnetically by interaction that appears to be transmitted through the electrons of interconnecting bonds. Such an effect is known as spin – spin coupling.

The signal for a proton is often split by spin – spin coupling with the proton on adjacent carbons. While this might seem to make things more complex, it provides valuable information about the structure of the molecule.

<u>**Example:**</u>

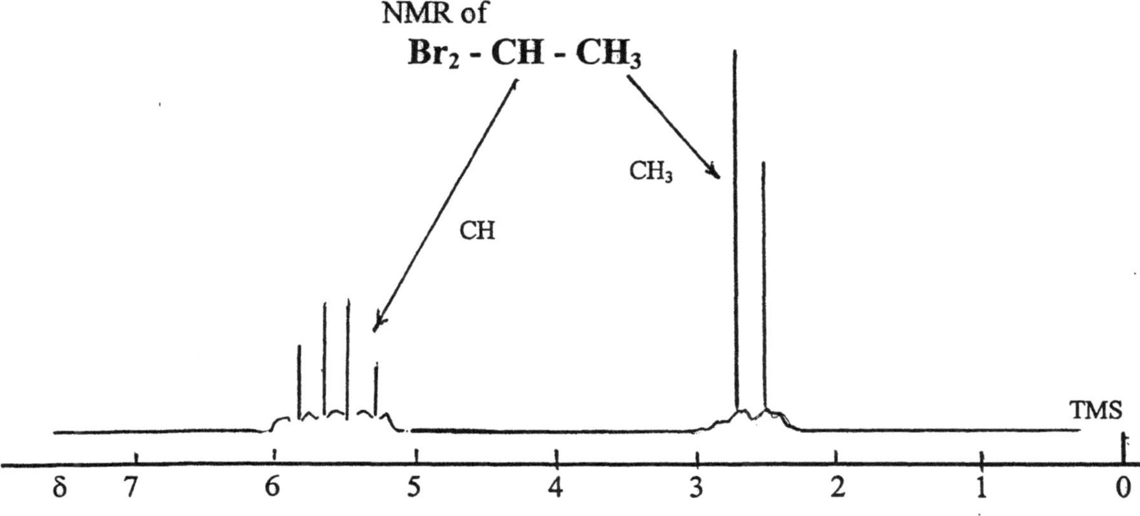

$$Br_2 - CH - CH_3$$

Two types of protons in 1,1 dibromoethane , so we expect to see NMR signals for two types of protons.

Why some protons show multiplets instead of single lines?

$$Br_2 - CH - CH_3$$

There are **two** ways to align the magnetic moment of this proton in an external magnetic field (aligned or opposed). The two possible alignments are:

$$Br_2 - CH - CH_3 \uparrow \qquad\qquad Br_2 - CH - CH_3 \downarrow$$

The field at the CH_3 group is affected by these spin states. About half the molecules have this spin aligned and about half opposed, so we see two slightly different methyl groups, i.e. **we see two lines from the methyl group because it is affected by two spin states of the single proton on the adjacent carbon.**

$$Br_2 - CH - CH_3$$

- The three protons in the CH_3 group could all have their spins aligned with the external field. This can be represented like this : ↑↑↑
- The lowest energy state for a CH_3 group has the spins of the three protons aligned with the field
- The next higher energy state has two aligned and one opposed as follows:

   ↑↑↓   **OR**   ↑↓↑   **OR**   ↓↑↑   (two up and one down)

   ↑↓↓   **OR**   ↓↑↓   **OR**   ↓↓↑   (one up and two down)

        and   ↓↓↓

The energy difference is small, it is **three times** more probable to find **2** up and **1** down as it is to find all spins up. Since **there are four spin states** for the three protons in a methyl group, we may find **the NMR signal from adjacent protons split into four lines.**

$$Br_2 - CH - CH_3$$
$$\uparrow$$

     4 lines for this proton are expected, i. e. the signal from this proton is split into 4 lines by the adjacent methyl group. These lines do not have the same intensity (intensities are in the ratio $1 : 3 : 3 : 1$).
Remember that the signal from the methyl group is split into two lines by the adjacent proton. Therefore, in the compound :

$$Br_2 - CH - CH_3$$

      4 lines   2 lines

   This means that the number of lines you see in the proton NMR depends on the number of adjacent protons.

<u>**Example :**</u>   In the compound $$Br - CH_2 - CH_3$$

      4 lines   3 lines

<table>
<tr><td colspan="2">The n + 1 rule</td></tr>
<tr><td>Adjacent Proton</td><td>Number of lines</td></tr>
<tr><td>1</td><td>2</td></tr>
<tr><td>2</td><td>3</td></tr>
<tr><td>3</td><td>4</td></tr>
<tr><td>n</td><td>n + 1</td></tr>
</table>

The n + 1 rule work best if the difference in chemical shift of the coupled protons is large. No splitting is observed when the chemical shifts of adjacent protons are the same.

NMR Spectrum of Ethyl Alcohol

$$CH_3 - CH_2 - OH$$

We expect to see three groups of lines in the proton NMR spectrum of the alcohol $CH_3 - CH_2 - OH$ because we have three different types of protons.

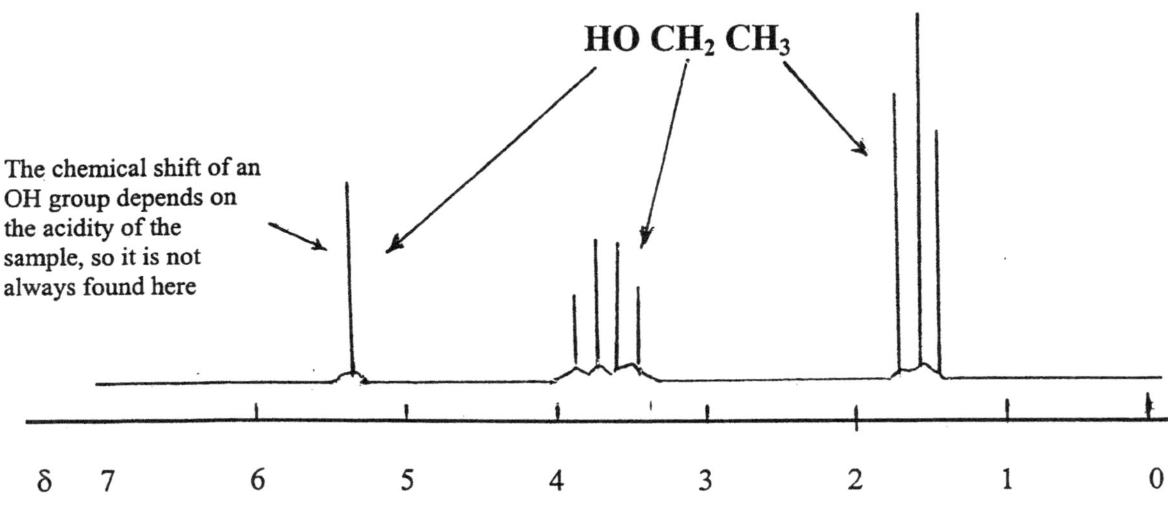

The CH_2 is shown as a group of 4 lines, i. e. the signal from the CH_2 protons split by the methyl group.

The spectrum does not show splitting of the OH proton because rapid exchange of the OH protons with other with other protons averages the states.

CHARACTERISTIC NMR SPECTRAL POSITIONS
FOR HYDROGEN IN ORGANIC STRUCTURES

By permission from Erno Mohacsi, J. of Chemical Education, 41, 38 (1964)

This table is useful for quick qualitative determination of proton spectrum lines by providing a tabulation of line positions obtained using tetramethylsilane as an internal reference. The listing has been kept as simple as possible for this purpose. The proton spectrum lines are arranged according to the chemical shift relative to tetramethylsilane and are given in values of τ and δ. The purpose of this table is to supplement tables available in standard references and to summarize information available in the literature.

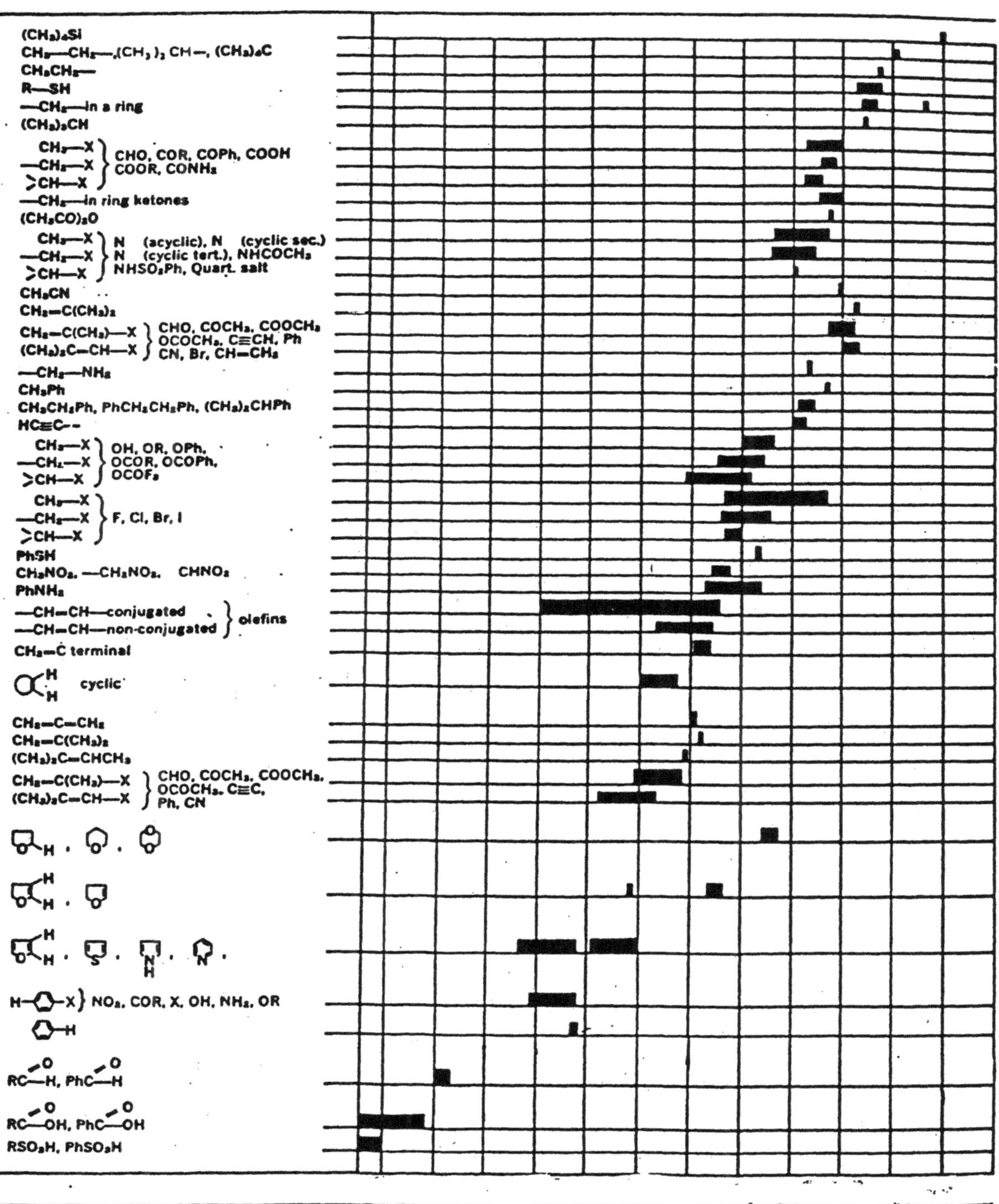

$$C_8H_7ClO$$

The uv spectrum of this compound is determined in 95% ethanol: λ_{max} 250 nm (log ϵ 3.2). The ir spectrum is obtained on a neat liquid sample.

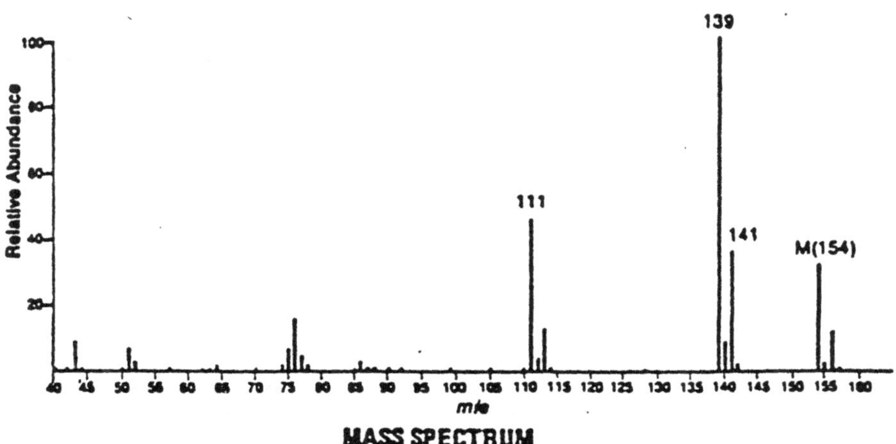

MASS SPECTRUM

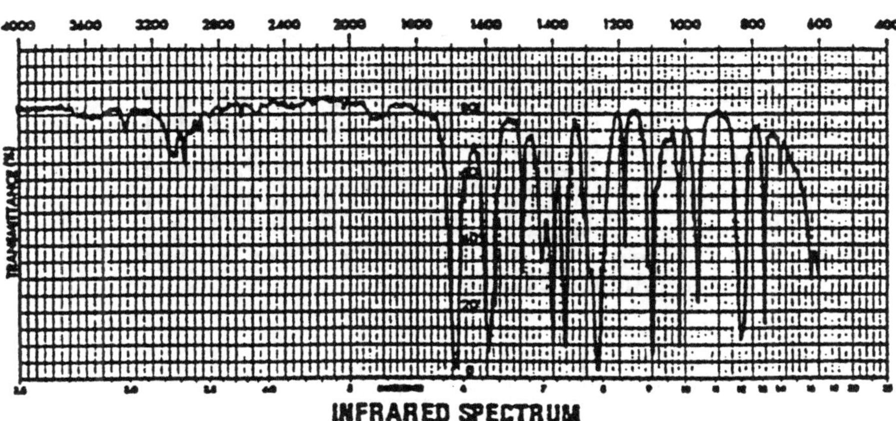

INFRARED SPECTRUM

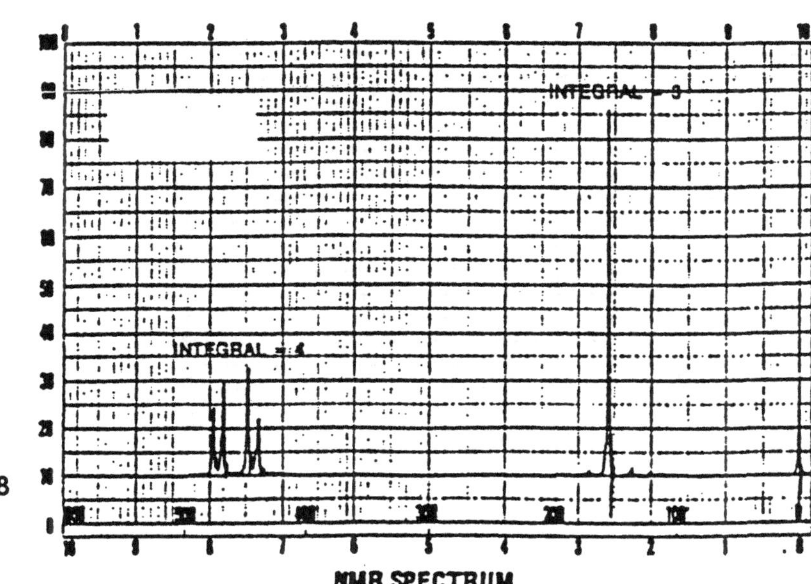

78

NMR SPECTRUM

$C_5H_{10}O$

The uv spectrum of this compound is determined in 95% ethanol: λ_{max} 290 nm (log ϵ 1.4). The ir spectrum is obtained on a neat liquid sample.

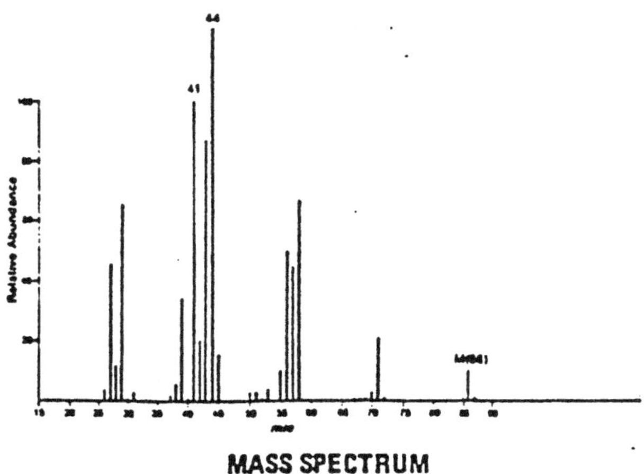

MASS SPECTRUM

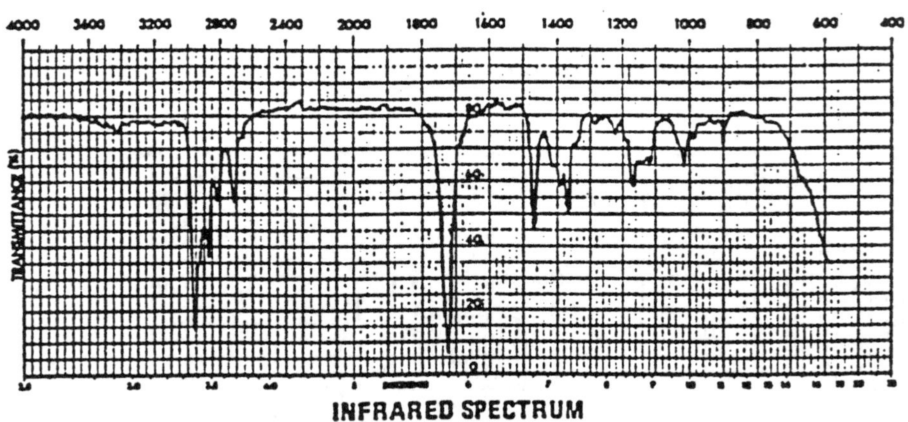

INFRARED SPECTRUM

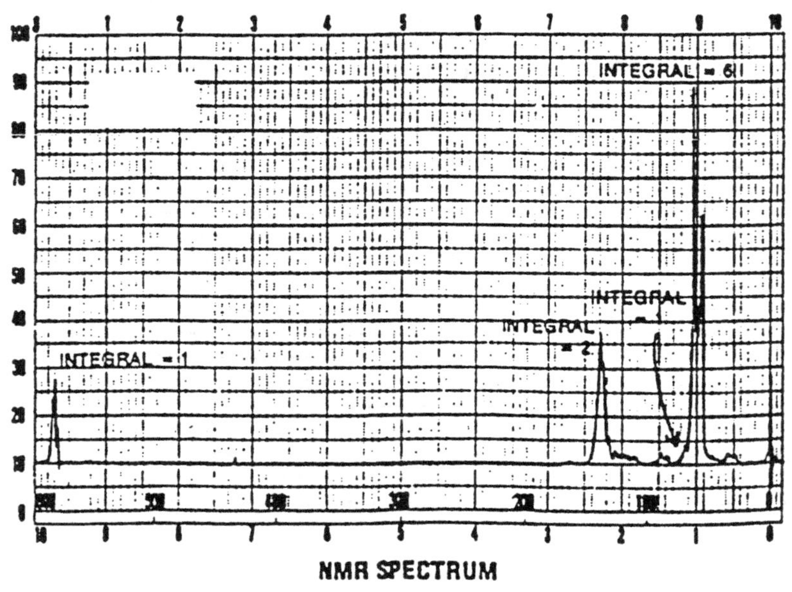

NMR SPECTRUM

$$C_8H_9NO_2$$

The uv spectrum of this compound shows an intense primary band, λ_{max} 251 nm (log ϵ > 4). The ir spectrum is obtained on a neat liquid sample.

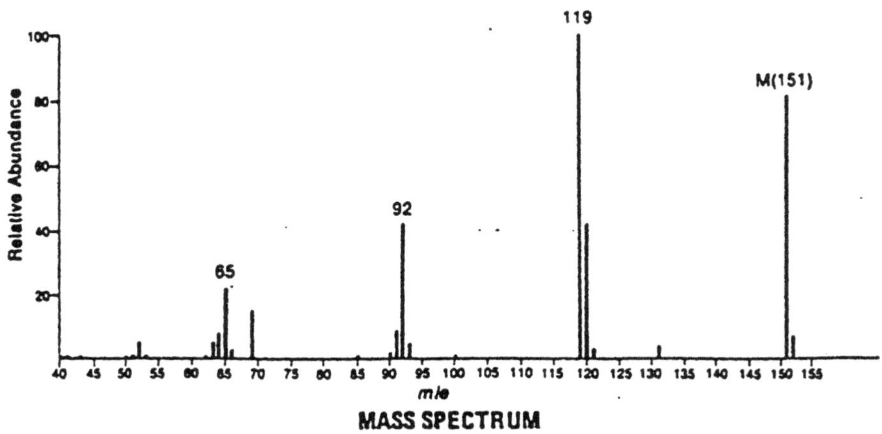

MASS SPECTRUM

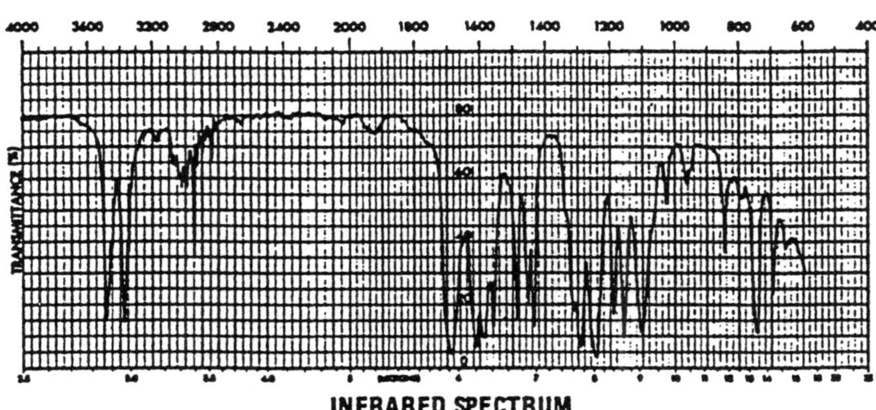

INFRARED SPECTRUM

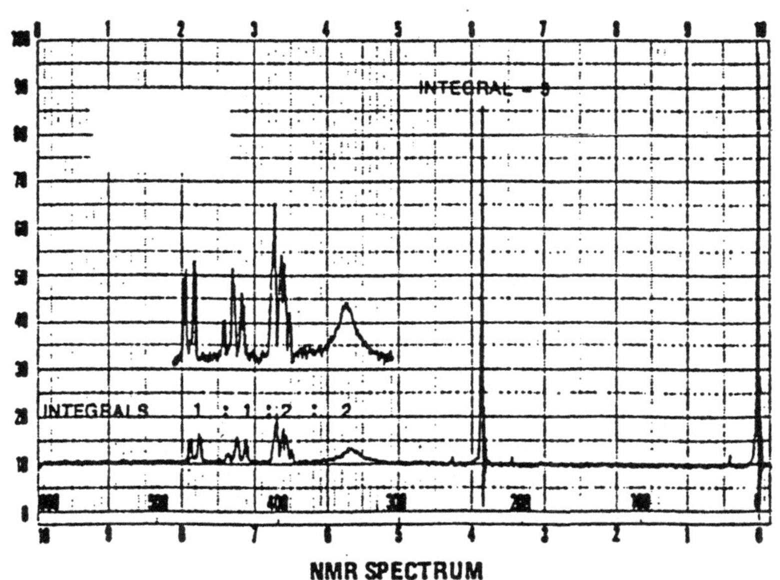

NMR SPECTRUM

$$C_6H_{15}N$$

The uv spectrum of this compound shows no maximum above 205 nm. The ir spectrum is obtained on a neat liquid sample.

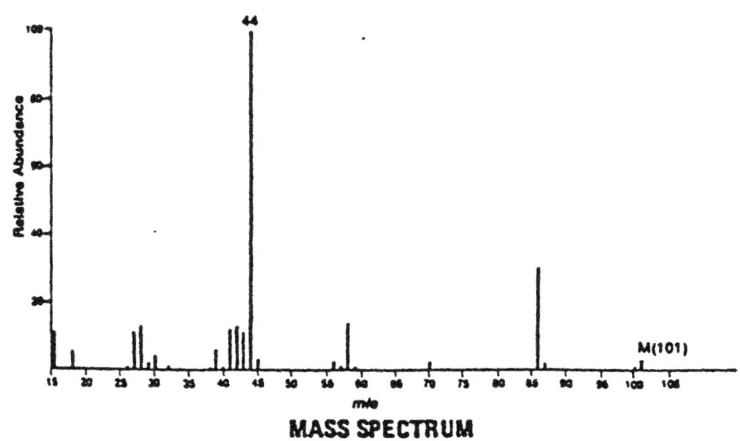

MASS SPECTRUM

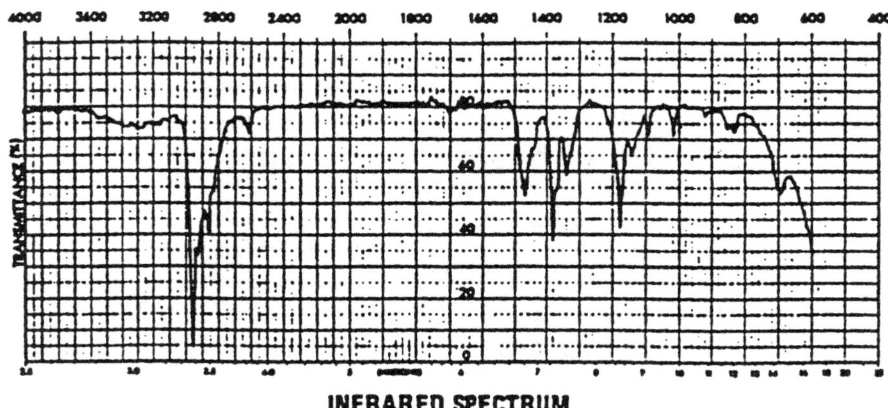

INFRARED SPECTRUM

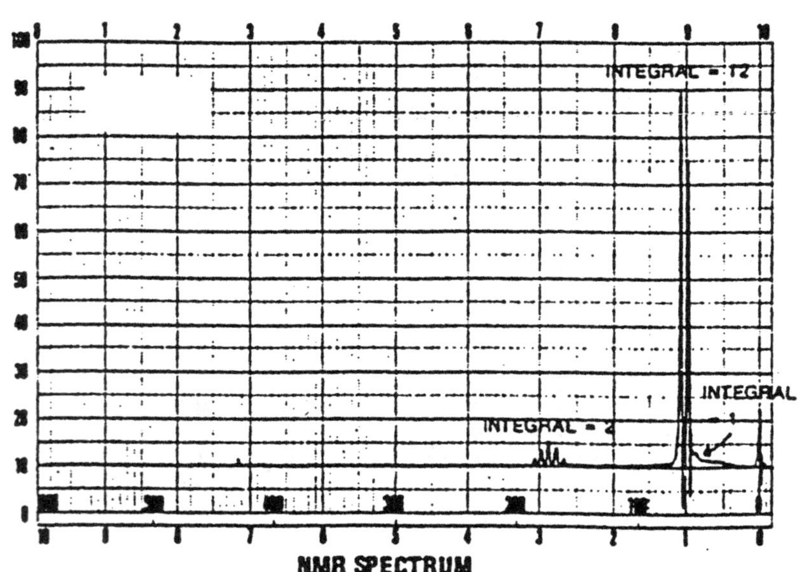

NMR SPECTRUM

Carbocation Rearrangements— Benzopinacolone

Background Reading
Material on the pinacol–pinacolone rearrangement in your lecture text

Timing
(30 min at beginning of period to isolate the benzopinacol prepared in Part **A** of Experiment 35; 1 h for the preparation of benzopinacolone)

A variety of organic reactions involve generation of carbocations followed by rearrangement (migration of a group) and the eventual formation of products with structures substantially different from those of the starting materials. An example is the rearrangement of benzopinacol to benzopinacolone.

$$
\underset{\text{Benzopinacol}}{\text{Ph}-\overset{\overset{\text{OH}}{|}}{\underset{\underset{\text{Ph}}{|}}{\text{C}}}-\overset{\overset{\text{OH}}{|}}{\underset{\underset{\text{Ph}}{|}}{\text{C}}}-\text{Ph}}
\quad\xrightarrow[\text{HOAc}]{\text{I}_2}\quad
\underset{\text{Benzopinacolone}}{\text{Ph}-\overset{\overset{\text{Ph}}{|}}{\underset{\underset{\text{Ph}}{|}}{\text{C}}}-\overset{\overset{\text{O}}{||}}{\text{C}}-\text{Ph}}
$$

The reaction proceeds by initial formation of a carbocation, followed by rearrangement of a phenyl group and loss of a proton to give the product. Evidence suggests that as the phenyl migrates, it forms a bridged intermediate called a **phenonium ion**.

Phenonium ion intermediate

1. Add one or two steps to the beginning of the above mechanism to show the function of the I_2 in the rearrangement of benzopinacol. Can you suggest any other reagent that might be used to effect the rearrangement?

2. Draw individual resonance-contributing structures showing positions to which the charge is delocalized in the phenonium ion.

3. Predict how the stability of the phenonium ion would change if it contained a p-CH$_3$O substituent; a p-NO$_2$ substituent.

CAUTION!

Glacial acetic acid is corrosive. Plastic gloves are recommended.

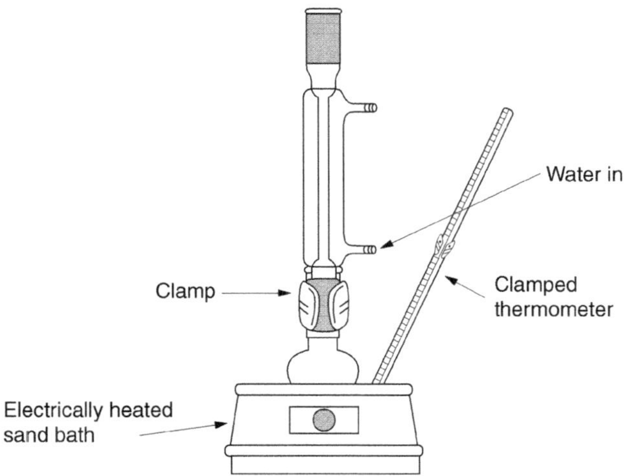

Figure E36.1 Apparatus for preparation of benzopinacolone

A. Benzopinacolone

Charge a small pear-shaped flask with 200 mg (0.546 mmol) of benzopinacol,[1] 1 mL of glacial acetic acid, and a tiny crystal (about 5 mg) of iodine. Attach a condenser and heat to reflux using a sand-bath (Figure E 36.1) for about 5 min. Allow the solution to cool. Benzopinacolone separates from solution. Add a little ethanol, mix thoroughly, and isolate the product by suction filtration on a small Hirsch funnel. Use additional ethanol to rinse the flask and to remove any iodine color from the product. Suck the benzopinacolone dry for a few minutes, weight it, and determine the yield and mp. The reported mp is 182–184°C. The yield should exceed 70%.

1. Either the commercial material or that prepared in Experiment 35A can be used.

EXERCISES

4. Because benzopinacol and benzopinacolone have nearly identical melting points, describe (without using spectroscopic methods) how you might prove that samples of the two compounds are different.

5. (**Optional; Spectroscopic**) Describe any important differences between benzopinacol and benzopinacolone in the IR, ^{1}H NMR and ^{13}C NMR spectra.

Carbocation Rearrangements

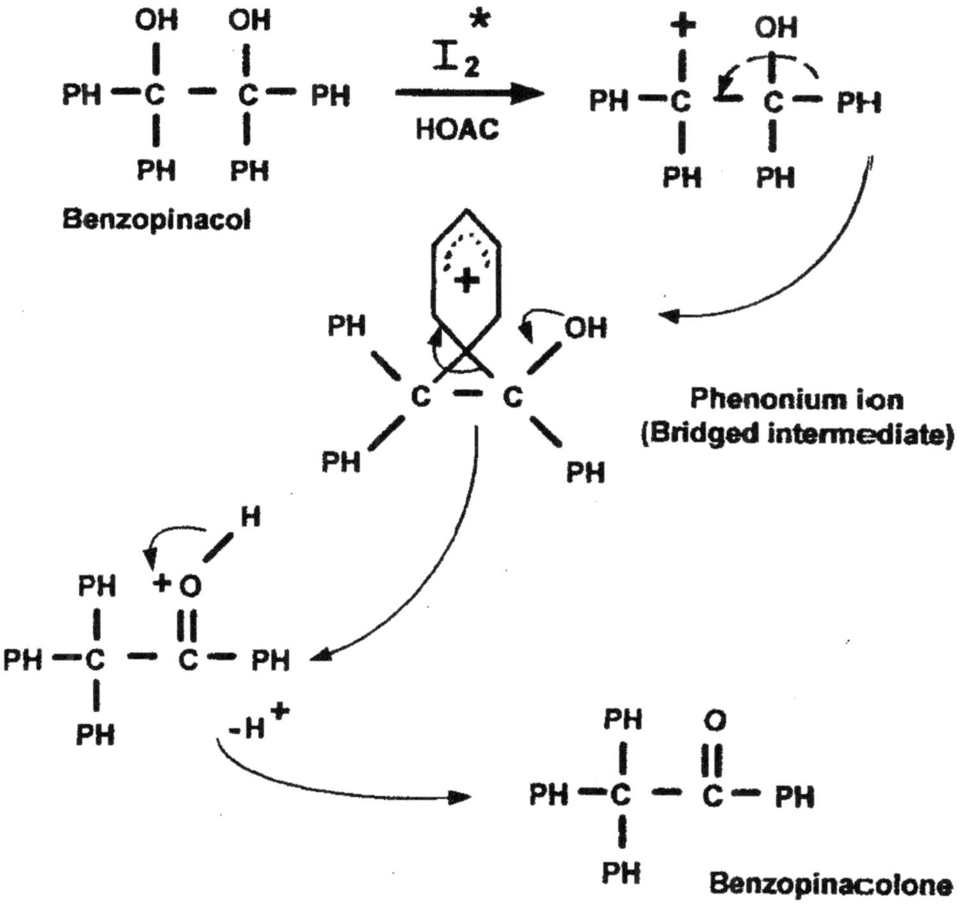

Ex1: *

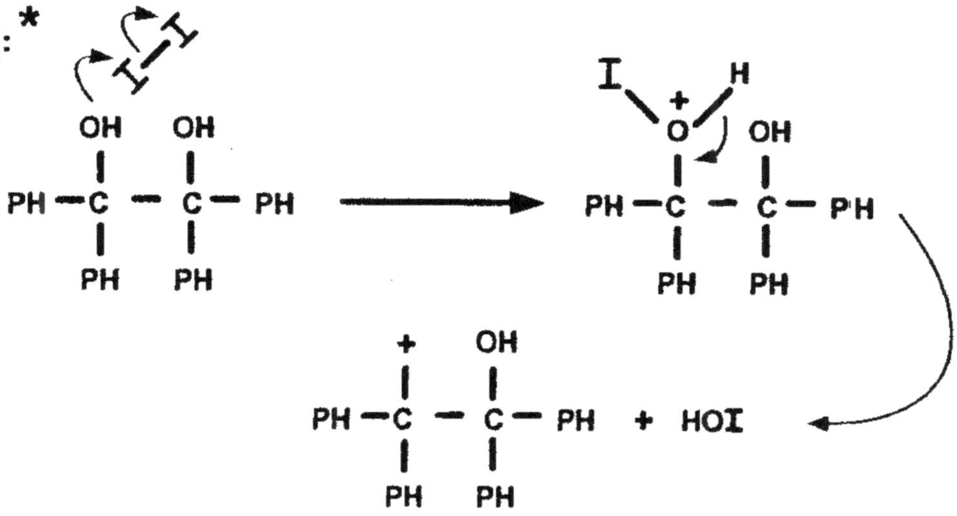

Methylcyclohexenes; Alcohol Dehydration

Background Reading
Material on alcohol dehydration in any organic textbook.

Timing
(3 h)

The starting material for this experiment is a commercial mixture of *cis-* and *trans-*2-methylcyclohexanol. Dehydration with concentrated phosphoric acid results in the formation of 1-methylcyclohexene as the major product together with a small amount of 3-methylcyclohexene, and, depending on reaction conditions, still smaller amounts of methylenecyclohexane.

$$ \text{2-Methylcyclohexanol} \quad (E+Z) \xrightarrow[\Delta,\ -H_2O]{H_3PO_4} \quad \text{Major product} $$

CAUTION!

Phosphoric acid is corrosive and should be handled with care. Plastic gloves are recommended.

A. Dehydration

With an automatic delivery pipet add 300 μL of 2-methylcyclohexanol (density = 0.93 g/mL, bp 163–166°C) to the boiler of a Hickman distillation unit, Figure E16.1. The thermometer should be arranged to measure the sand bath temperature. Then add a boiling chip and 75 μL of concentrated (85%) phosphoric acid. Swirl the still gently to mix the reactants and clamp the still so that the boiler tube (bottom) protrudes into the sand. Heat the alcohol–acid mixture until product begins to steam distill. The temperature of the sand bath should be about 140°C. Cool the cloth wrapped around the collection ring with acetone and water. Continue collecting a cloudy mixture of product and water until distillation stops. Transfer the product from the collection ring of the still to a small vial with a Pasteur pipet, add about 50 mg (estimated) of sodium bicarbonate and about 50 mg of anhydrous sodium sulfate, swirl the suspension to ensure that all water droplets come into contact with the drying agent, and allow the capped vial to stand for about 5–10 min. (Use a cap with a Teflon liner or use a piece of polyethylene film to keep the liquid out of contact with the cap liner.)

Transfer the clear liquid with a filter pipet to a weighed vial and calculate your yield. (Your yield should be at least 70% in this apparatus.) Obtain a boiling point on a small sample with the special procedure described in Section 3.2 and Figure 3.8. (Store your sample in a vial with a Teflon liner in the cap.)

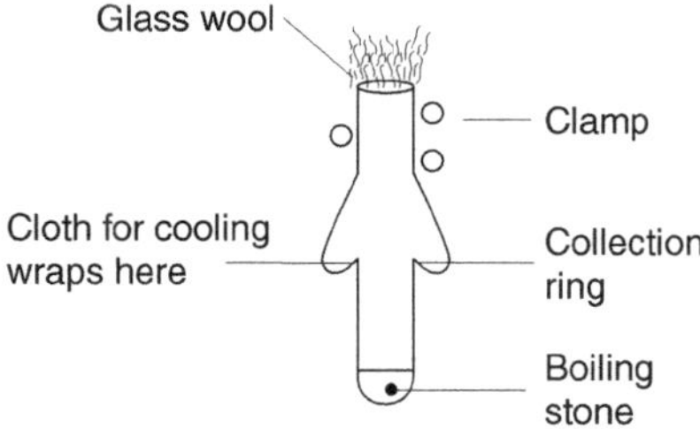

Figure E16.1 Hickman distillation unit

B. Baeyer Test

Use one small drop of your sample to run the following Baeyer test. Prepare a violet solution from a tiny (barely visible) amount of potassium permanganate and about 2 mL of acetone in a small test tube. Add a few drops of the alkene product and note what happens. What is the brown precipitate that forms?

C. Gas Chromatography

Inject a 1–2-μL sample of your product onto an FFAP gas chromatography column at an oven temperature no greater than 90°C. Assign the two major peaks to the appropriate alkene structure. Assume that the response factors are identical for the two products and determine the approximate mol ratio of major to minor alkene.

NOTE

The relative amounts of 1-methylcyclohexene and 3-methylcyclohexene will vary depending on the individual reaction conditions and extent of recovery of product alkenes that have different boiling points.

D. Bromine Test

Put one drop of your alkene sample in about 0.5 mL of methylene chloride and add a drop of a freshly prepared[1] 2% (by volume) solution of bromine in methylene chloride. Carefully note the results and describe the reaction with an appropriate equation for each of the two major alkenes present. Try the same test with both trichloroethylene and toluene (methylbenzene). Describe the results. Allow the test mixtures to stand for as long as 20–30 min.

EXERCISES

1. Write a mechanism that describes the formation of the two more important products, 1-methylcyclohexene and 3-methylcyclohexene.

2. Give a mechanism for the formation of the minor product, methylenecyclohexane (which is sometimes not observed at all). Why is methylenecyclohexane a more likely minor product than 4-methylcyclohexene?

1. Bromine does react slowly with methylene chloride to liberate HBr over a period of days.

3. Predict which three alkenes will be formed in the dehydration of 3-methyl-3-pentanol.

4. The hydration of an alkene to give an alcohol can be accomplished in dilute aqueous acid by a mechanism that is the reverse of that for alcohol dehydration. Start with isobutylene and illustrate how a hydration reaction would occur.

5. The bromine test (Part **D**) is often used as an indication of unsaturation (double and triple bonds). Explain why your results for trichloroethylene and toluene were different than for the simple alkene products from your reaction mixture from Part **A**.

Alcohol Dehydration

Dehydration of alcohols is a reaction in which elements of water are removed from alcohol.

Example: Dehydration of cyclohexanol to cyclohexene.

$$\text{cyclohexanol} \xrightarrow{\;H_3PO_4\;} \text{cyclohexene} + H_2O$$

cyclohexanol   **cyclohexene**

Heat and Lewis acids such as alumina (Al_2O_3) can be used to catalyze the reaction.

Mechanism:

The reaction occurs by three- step mechanism involving a carbocation intermediate.

1. In the first step, the OH group of the alcohol accepts a proton from the catalyzing acid in a Bronsted acid – base reaction.

2. The carbon-oxygen bond of the alcohol breaks in a Lewis acid- base dissociation to give water and a carbocation.

a carbocation

3. Water, the conjugate base of the catalyzing acid H_3O^+ removes a proton from the carbocation in another Bronsted acid-base reaction.

H   ÖH₂
any one of the four
β-hydrogens
+ H—ÖH₂

This Step generates the alkene product and regenerate the catalyzing acid H_3O^+.

Procedure:

1. Use the micro scale set, place 1 ml of 2-methylcyclohexanol in the flask, then add a boiling chip and 0.3 ml of concentrated phosphoric acid, swirl the flask to mix the reactants.

2. Heat the mixture gently at reflux until product begins to steam. (the product is formed at this point. We are not separating the product for safety.

3. Cool and transfer the product to a small test tube. Add about 0.5g sodium bicarbonate and 0.2g of anhydrous sodium sulfate. Swirl and allow the test tube to stand for about 5 minutes

4. Transfer the clear liquid to two small test tubes and perform Baeyer test on one sample and bromine test on the second as described on page 365.

Discard products of all tests in the waste container provided in Lab.

Hydroboration-Oxidation of 1-Hexene[1]

Background Reading
Study the discussion on hydroboration of alkenes in your lecture text.

Timing
(5 h)

The most widely used reaction to achieve anti-Markovnikov addition of water to an alkene is hydroboration-oxidation. Not only can this reaction be applied to a variety of structurally different alkenes, but yields are very good and the addition is both highly regioselective and stereoselective (syn-addition).

The reaction sequence normally employed involves the initial treatment of the alkene in dry tetrahydrofuran (THF) with borane-tetrahydrofuran (BH$_3$·THF) under oxygen-free conditions (achieved with an atmosphere of nitrogen or argon).

$$BH_3 \cdot THF + 3\ RCH{=}CH_2 \longrightarrow THF \cdot B(CH_2CH_2R)_3$$

The resulting organoborane is then oxidized under basic conditions with 30% hydrogen peroxide.

$$THF \cdot B(CH_2CH_2R)_3 + 3\ HOOH \longrightarrow 3\ RCH_2CH_2OH + B(OH)_3 + THF$$

In this experiment the borane is generated *in situ* (within the reaction mixture) by the oxidation of borohydride with iodine in the presence of 1-hexene.

$$2\ THF + 2\ BH_4^- + I_2 \longrightarrow 2\ BH_3 \cdot THF + H_2{\uparrow} + 2I^-$$

$$BH_3 \cdot THF + 3\ BuCH{=}CH_2 \longrightarrow THF \cdot B(CH_2CH_2CH_2CH_2CH_2CH_3)_3$$

Because the borane reacts quickly as it is produced and evolves hydrogen, there is no need for an inert atmosphere.

In addition, the somewhat dangerous 30% hydrogen peroxide often used for the oxidative work-up has been replaced with the safer and easier to handle sodium perborate tetrahydrate, NaBO$_3$·4 H$_2$O (more accurately written as NaBO$_2$·H$_2$O$_2$·3 H$_2$O). Note that the latter reagent is a source of hydrogen peroxide in aqueous solution.

EXERCISES

1. Write a mechanism for the addition of borane to 1-hexene. Draw the structure of a transition state for this process that helps to explain the dominant regiochemistry that is observed.

2. Write a balanced ionic equation for the oxidation of trihexylborane to 1-hexanol with hydroperoxide, HOO$^-$, in a basic medium.

---

1. Adapted from G. W. Kabalka, P. P. Wadgaonkar, and N. Chatia, *J. Chem. Educ.*, **67**, 975 (1990).

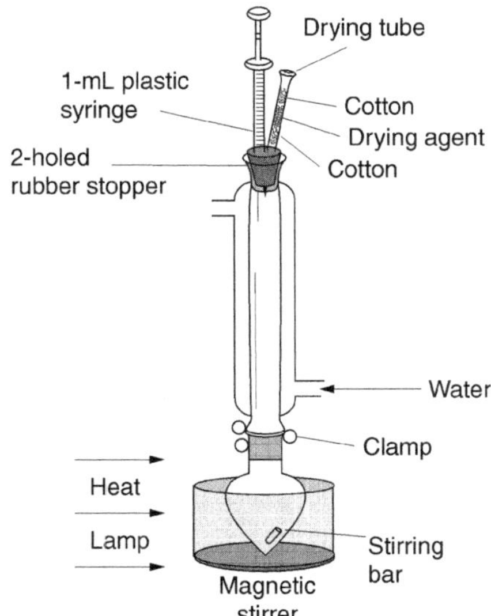

Figure E39.1 Apparatus for hydroboration

A. Hydroboration of 1-Hexene

The apparatus consists of a 25-mL pear-shaped flask with a magnetic stirring bar, and a reflux condenser with a two-holed rubber stopper containing a drying tube made from a medicine dropper and a plastic 2.5-mL syringe[2] with a 40-mm-long, #25 needle[3] (Figure E39.1) The flask is placed in a shallow dish (such as a small crystallizing dish) on the top of the magnetic stirrer. The flask, condenser, drying tube, and plastic syringe should be oven dried prior to use.

Remove the flask from the oven, allow it to cool, and weigh directly into it 63 mg (1.66 mmol) of sodium borohydride followed by the addition of 4 mL of dry tetrahydrofuran,[4] and 500 μL (337 mg, 4.00 mmol) of 1-hexene. Attach the warm condenser with stopper, drying tube, and syringe, and cool the reaction mixture with ice (crystallizing dish). Turn on the cooling water in the condenser.

Weigh 240 mg (0.946 mmol) of iodine into a 1-dram vial, add about 2 mL of dry THF, cap the vial, and swirl it until solution is complete. Draw the iodine solution into the syringe, and add it slowly (10 min) to the stirred, cold reaction solution.[5] (A small amount of hydrogen will be evolved.) Remove the cold bath and allow the reaction mixture to stir for 45 min.[6]

Carefully decompose the excess borohydride by the addition of about 2 mL of water in 3–4 portions. (Hydrogen will be evolved.)

B. 1-Hexanol

To oxidize the borane from Part **A**, add 415 μL of 3 M sodium hydroxide solution followed by 765 mg (4.97 mmol) of sodium perborate tetrahydrate.[7] After a mild exothermic reaction begins (10–15 min; normally accompanied by a color change to yellow), cool the mixture with ice water to avoid the formation of by-products.[8] Oxidation should be complete after an additional 30 min.

Add about 500 μL of saturated sodium thiosulfate with stirring to destroy excess oxidizing agent[9] followed by 2 mL of hexane (Skelly B), and just enough anhydrous potassium carbonate to saturate the aqueous layer. If the yellow-brown iodine color

2. The polyethylene syringe will become stained from the iodine, but can be cleaned with acetone, dried in an oven, and reused for this experiment.
3. The exact size of the needle is not critical.

4. Reagent-grade THF should be dried for at least several hours over anhydrous calcium sulfate.

5. A stoichiometric excess of borohydride compensates for the loss of some borohydride by reaction with moisture and the lack of a quantitative reaction with the iodine. The iodine color will persist for 20–30 min of stirring after the cold bath is removed.
6. To avoid corrosion of the syringe needle by prolonged contact with iodine, replace the syringe with a short piece of stirring rod. Rinse the syringe and needle with acetone as soon as possible and set it aside for return to the stockroom or instructor.
7. Because the oxidizing agent is moisture sensitive, it should be stored in a tightly closed container in a cool location when not in use.
8. The yellow iodine color may appear during the oxidation but will not interfere with the oxidation of the organoborane.
9. The solution should be colorless.

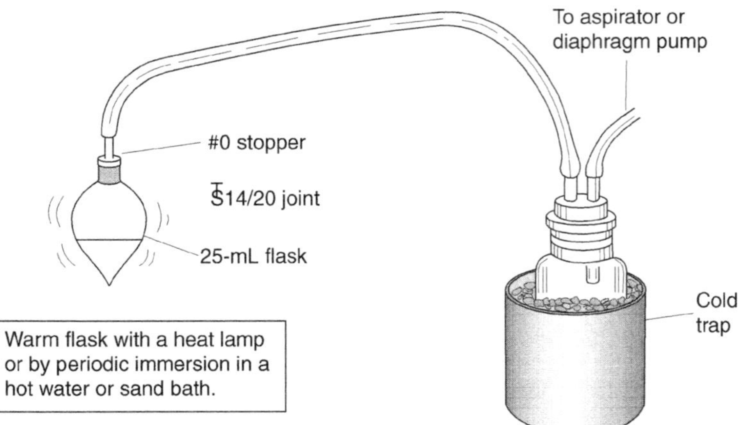

Figure E39.2 Method for solvent evaporation

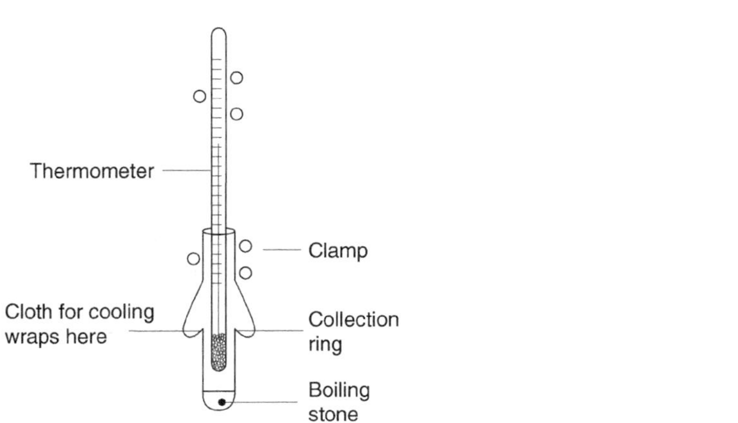

Figure E39.3 Distillation of 1-hexanol (heat with sand bath)

reappears, add a few drops more of the thiosulfate solution and continue to stir until the iodine is reduced to colorless iodide. Allow the layers to settle and carefully draw off the lower aqueous layer with a Pasteur pipet. Dry the organic layer over a small amount of anhydrous Na_2SO_4, transfer the solution with a filter pipet to a small, dry, pear-shaped flask, and evaporate the excess solvent at room temperature by the method shown in Figure E39.2 to obtain crude and often lightly colored 1-hexanol (liquid).

Transfer the crude product with a Pasteur pipet into a Hickman still, add a small chunk of anhydrous calcium sulfate as a boiling chip, and distill all of the 1-hexanol using a sand bath as shown in Figure E39.3. Do not cool the still until the thermometer exceeds 100°C. The final product is a colorless liquid, bp 158°C. Although the actual yield of alcohol at the completion of the oxidation step exceeds 90%, unavoidable losses result in a yield of isolated and purified product in the range of 30–40%.

3. What other product might be expected if the reaction were less than 100% regiospecific? How would you check for the presence of that product[10] in the crude reaction mixture? Would the two regioisomeric hexanols separate cleanly during the final distillation?

10. The literature indicates that about 6% of the minor regioisomer forms in this reaction.

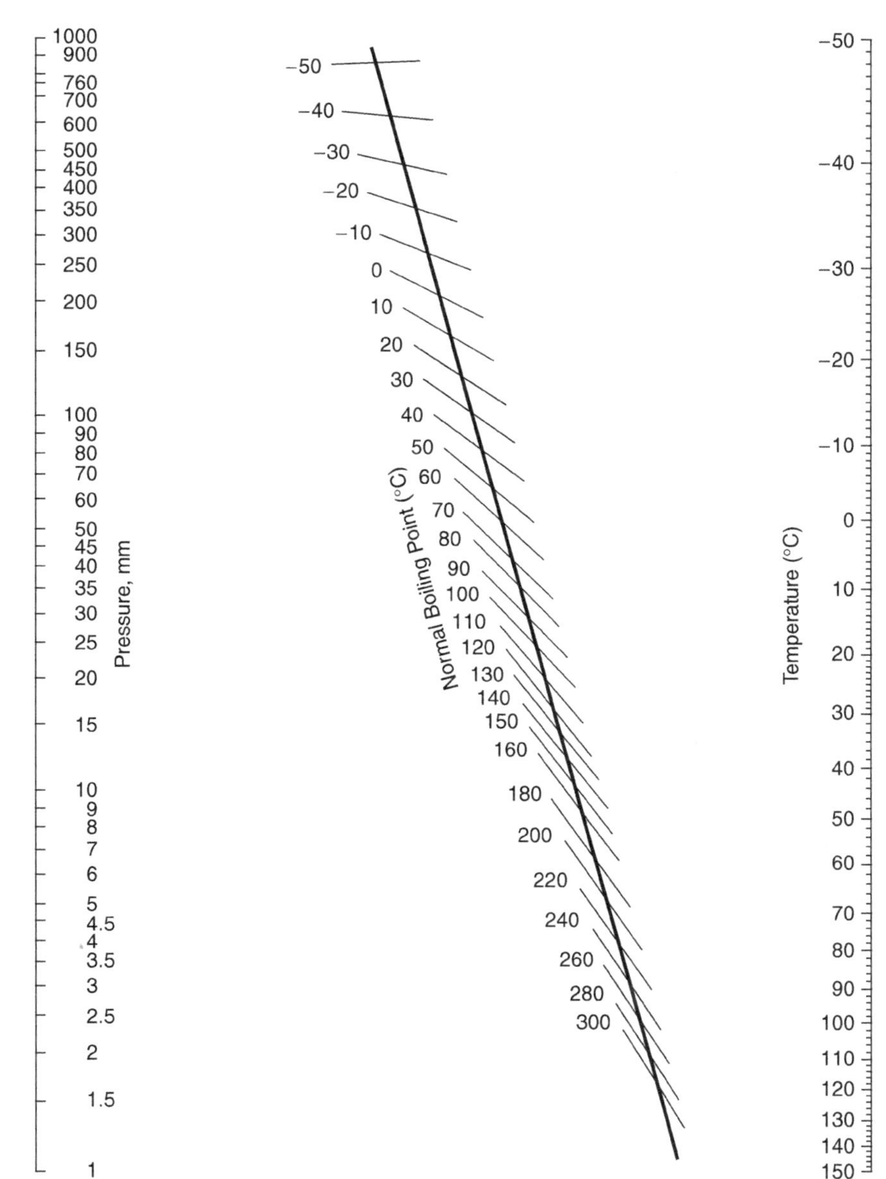

Simplified Vapor Pressure–Temperature Nomograph
(Source: S. B. Lippincott and M. M. Lyman, *Ind. Eng. Chem.*, **38,** 320 (1946).

Decreasing Compound Polarity

Order	Structure	Name
First	RCO_2H	Acids
Second	ArOH	Phenols
Third	ROH	Alcohols
Fourth	RNH_2	Amines
Fifth	RCO_2R	Esters
Sixth	RCOR	Ketones
Seventh	ROR	Ethers
Eighth	RX	Alkyl halides
Ninth	C=C—C=C	Conjugated dienes,
	ArH	Aromatic hydrocarbons
Tenth	C=C, C≡C	Unsaturated hydrocarbons
Eleventh	RH	Saturated hydrocarbons

Efficiency of Drying Agents for Organic Liquids

K_P	Number of Extractions (Vol. of phase 2 = Vol. of phase 1)				Number of Extractions (Vol. of phase 2 = $\frac{1}{4}$ Vol. of phase 1)			
	1	2	3	4	1	2	3	4
2	66.7	88.9	96.3	98.8	33.3	55.5	70.3	80.2
3	75.0	93.8	98.4	99.6	42.9	67.4	81.4	89.4
4	80.0	96.0	99.8	99.8	50.0	75.0	87.5	93.8

CHEMISTRY 204

CHEMISTRY 204
LABORATORY SCHEDULE

WEEK	EXP. #	TITLE
1	EXP. 30	Heat of Reaction: Measurement of Resonance Energy
2	EXP. 29	Kinetic and Thermodynamic Reaction Conditions
3	EXP. 15	Kinetic Investigation of Unimolecular Solvolysis
4	EXP. 43	Fisher Esterification
5	EXP. 48	9 – Benzal and 9 –Benzylfluorene Aldol and Cannizaro Reactions
6	EXP. 26	Oxidation of (-) Borneol to (-) Camphor
7	EXP. 34	4-methylbenzophenone Friedle – Crafts Acylation
8	EXP. 41	Electronic Effects of Substituents in Reactions: Acidity Constant Determinatin
9	EXP. 51	Synthetic Dyes
10	EXP. 51	Synthetic Dyes Continued
11	EXP. 40	Synthetic Organic Polymers
12	EXP. 54	Qualitative Organic Analysis: Classification Test
13	EXP. 55	Qualitative Organic Unknowns
14	EXP. *	Carbohydrate Lab

LABORATORY MAKE UP AT THE DISCRETION OF THE LABORATORY INSTRUCTOR.

* Original lab

Heat of Reaction; Measurement of Resonance Energy

A multiple bond such as is found in a carbonyl group ($C = O$) becomes stabilized when it is connected by a single bond to (conjugated with) another multiple bond. Such stabilization is associated with an increased distribution (delocalization) of the π-electrons, which is possible in a conjugated system relative to a nonconjugated system. The decrease in energy on going from a nonconjugated to a conjugated multiple bond is known as **resonance**, or **delocalization energy**, and is expressed in kcal/mol or kjoules/mol[1].

$$CH_3 - CH = CH - \overset{\displaystyle O}{\overset{\|}{C}} - CH_3 \qquad CH_2 = CH - CH_2 - \overset{\displaystyle O}{\overset{\|}{C}} - CH_3$$

A conjugated enone   A nonconjugated enone

Thermochemical techniques provide a convenient means of estimating the value of resonance energy by comparing heats of reaction for two closely related processes. In this experiment you will measure the total heat changes for the oxidation of α-methylbenzyl alcohol and isopropyl alcohol to the corresponding ketones by means of a simple calorimeter (shown in Figure E30.1).[2]

α-Methylbenzyl alcohol   Acetophenone

Isopropyl alcohol   Acetone

  In both reactions the starting material is a secondary alcohol. But of the two products, only acetophenone contains a carbonyl group conjugated with an aromatic ring. The difference in the total heat changes of the two reactions represents a combination of several effects. First is the electron delocalization or resonance, which is possible in the carbonyl group of acetophenone but not that of acetone.

Background Reading
Material on resonance or delocalization energy in any organic textbook. Students should be familiar with a method for balancing oxidation–reduction equations.

Timing
(3 h total; Part **A**; 1.5 h; Part **B**; 1.5 h)

1. Because resonance energy represents a stabilizing effect, it is understood to be a negative quantity. However, the minus sign is frequently omitted.

2. Construction details are given in the Instructor's Manual.

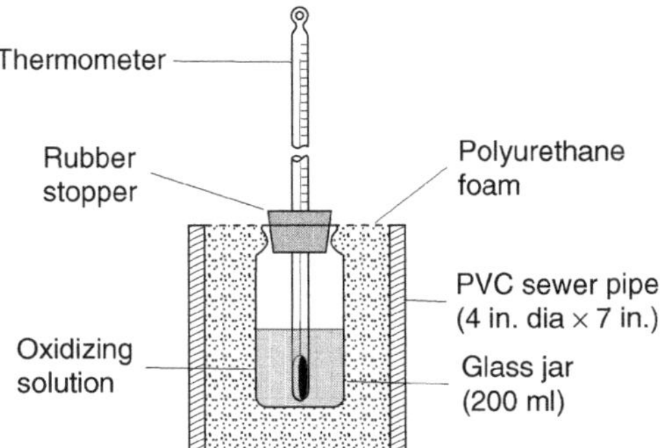

Figure E30.1   Calorimeter for heat of reaction measurements

Second (and usually included in the numbers quoted for resonance energy) are the differences in certain sigma-bond energies due to changes in hybridization. For example, the C_1—C_2 bond of isopropyl alcohol represents the overlap of two sp^3 orbitals, whereas in acetone it involves the overlap of an sp^3 and an sp^2 orbital. A similar comparison for the bond between the hydroxyl-bearing carbon atom and the ring-carbon atom of α-methylbenzyl alcohol with the corresponding bond in acetophenone shows that in α-methylbenzyl alcohol we have overlap of sp^2 and sp^3 orbitals, whereas in acetophenone we have the overlap of sp^2 and sp^2 orbitals. Thus, changes in the energy of this sigma-bond should not be identical for the two reactions being compared. However, for the determination of the resonance energy of the carbonyl group in acetophenone, we will assume that sigma-bond energy effects are small.

Last are effects caused by heats of solution. Because we begin the reaction by dissolving a starting alcohol in the reaction medium and because this alcohol is removed from solution by conversion to product ketone, the net effect from the heat of solution of the starting material is nil. However, the difference in the heats of solution for the organic products of the two reactions does contribute to the total heat changes being compared. Fortunately, this difference for acetophenone and acetone is small and can be neglected.

The evolution of a specific amount of heat during the reaction will increase the temperature inside different calorimeters by different amounts. Therefore, each calorimeter must be calibrated to determine how many calories of evolved heat will raise the temperature 1°C. One simple method is to carry out a reaction for which the total heat change is known under a specified set of reaction conditions.[3] In this experiment you will oxidize a weighed amount of hydroquinone to quinone and measure the change in temperature.

3. A better procedure (but requiring more elaborate equipment) is to pass a current through a small piece of resistance wire immersed in the solution within the calorimeter. The number of calories to raise the temperature of the solution 1°C is given by $Vit/4.18\Delta T$, in which V = voltage, i = current (amps), t = time (sec), and ΔT = total temperature change.

$$\text{HO}-\!\!\!\bigcirc\!\!\!-\text{OH} \xrightarrow[\text{Acetone, H}_2\text{O}]{\text{CrO}_3,\ \text{H}^+} \text{O}=\!\!\!\bigcirc\!\!\!=\text{O}$$

Hydroquinone   Quinone

1. Knowing that the product of reduction of CrO_3 is Cr^{3+}, balance the equations for the oxidation of α-methylbenzyl alcohol, isopropyl alcohol, and hydroquinone. How many mol of excess oxidant are used in Parts **A** and **B**?

2. For a calorimeter to be useful, it should be operated under as nearly adiabatic conditions as possible. Explain what is meant by that statement. What aspects

of the design of the calorimeter help to ensure this condition? Besides convenience, why is it more desirable to carry out the reaction in your calorimeter over a period of 10–15 min rather than 3–4 h?

3. ΔH for the hydrogenation of two mol of ethylene is −65.6 kcal/mol and for the complete hydrogenation of one mol of 1,3-butadiene is −57.1 kcal/mol. Account for the difference between these two numbers.

CAUTION!

Chromium trioxide or chromium(VI) oxide should be handled with gloves in a hood. All chromium compounds should be considered potentially carcinogenic when particles are ingested or inhaled. Chromium trioxide is a severe skin irritant.

WASTE DISPOSAL NOTE

Solutions containing chromium waste should be placed in suitably labeled containers.

NOTE

The thermometer should include the temperature range 20–40°C and should be graduated with 0.2°C or 0.1°C divisions. Temperature readings should always be estimated to the nearest 0.05°C.

A. Calorimeter Calibration

Set up the calorimeter as shown in Figure E30.1.

Add 30 mL of water to a 150-mL beaker and then, while stirring, slowly add 3 mL of concentrated sulfuric acid. Quickly (to avoid absorption of moisture from the air) weigh approximately 1 g (10 mmol) of chromium(VI) oxide onto a piece of glazed paper and dissolve it in the stirred sulfuric acid solution. Then add 70 mL of acetone and cool the resulting oxidizing solution to approximately 25–27°C with an ice- or cold-water bath. Pour the aqueous acetone solution into the calorimeter and close it with the stopper containing the thermometer.

NOTE

As you add the hydroquinone to the calorimeter in the next part of the experiment, be careful not to get it on the upper inside of the glass vessel or on the upper part of the thermometer or stopper, where it might not come into contact with the oxidizing solution.

Place about 700 mg of hydroquinone into a small screw-capped vial and weigh it to the nearest mg. Gently swirl the closed calorimeter to stabilize the temperature, and then record the temperature. Next, open the screw-capped vial and, while holding the stopper and thermometer to one side (as shown in Figure E30.2), add the hydroquinone, quickly replace the stopper, note the time with the sweep second hand of a watch or clock, and gently swirl the calorimeter.

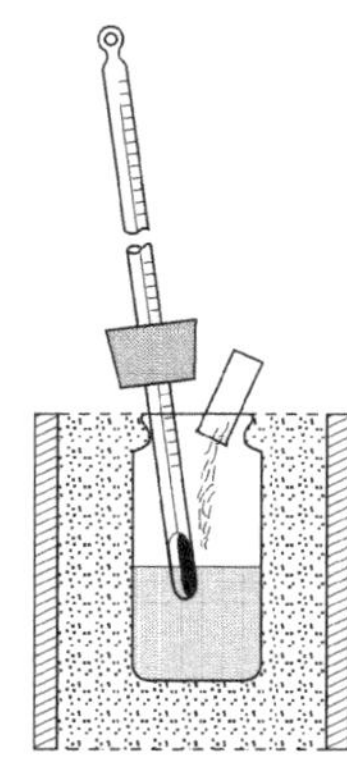

Figure E30.2 Addition of hydroquinone to the calorimeter

  Experiment 30 Heat of Reaction; Measurement of Resonance Energy

Continue to swirl the calorimeter and record the temperature every 0.5 min for about 3 min or until the temperature becomes constant. Next, replace the screw cap on the vial and weigh it to the nearest mg to determine how much hydroquinone was oxidized.

Repeat the experiment at least twice, each time with a fresh batch of oxidant.

EXERCISE

4. Given the fact that the heat change for the oxidation of hydroquinone to quinone by chromium trioxide in aqueous acidic acetone is −34.5 kcal/mol, calculate the average number of calories needed to raise the temperature of your calorimeter 1°. (Note that the figure −34.5 consists of both the heat of reaction of the hydroquinone and the heat of solution of the quinone.)

B. Resonance Energy

Prepare the oxidizing solution as described in Part **A**, but with 2 g (20 mmol) of chromium(VI) oxide instead of 1 g. Pour the cooled solution into the calorimeter. Gently swirl the closed calorimeter to stabilize the temperature, and then record the temperature. Carefully loosen the rubber stopper. Using a rubber bulb, carefully fill a graduated 1-mL pipet with α-methylbenzyl alcohol.

While holding the stopper and thermometer to one side, pipet 0.958 mL (8 mmol) of the alcohol directly into the oxidizing solution, quickly replace the stopper, note the time with the sweep second hand of a watch or clock, and gently swirl the calorimeter. Record the temperature every minute for 5 min or until the temperature becomes constant. Repeat the experiment at least once with a fresh solution of oxidant. Successive runs should result in temperature changes that fall within the range ±0.1°C.

Carry out at least two determinations for the heat change for the oxidation of 0.613 mL (8 mmol) of isopropyl alcohol. In this instance, temperature will have to be recorded over a span of approximately 15 min. For all runs in Part **B**, prepare a plot of temperature versus time.

EXERCISES

5. Using the calibration data determined in Part **A** together with data from Part **B**, calculate the average heat changes in kcal/mol for the oxidation of α-methylbenzyl alcohol and isopropyl alcohol. Assuming that the heat of solution effects and effects due to changes in σ-bond energy are small, what is the approximate resonance energy for the carbonyl group in acetophenone?

6. The heats of solution from acetone and acetophenone are −0.8 and −1.7 kcal/mol, respectively. Use these data to correct the resonance energy value determined in Exercise 5.

7. What conclusions can be drawn about the relative rates of oxidation of isopropyl alcohol and α-methylbenzyl alcohol by chromium(VI) oxide?

8. Would you expect the ΔH for oxidation of α-methyl-p-hydroxybenzyl alcohol to be more or less negative than that for α-methylbenzyl alcohol? Explain.

Kinetic and Thermodynamic Reaction Conditions

Background Reading
Material on semicarbazone formation and on kinetic and thermodynamic control of reactions.

Timing
(5 h; furfural should be purified in advance by the instructor)

Consider a reaction in which two products can arise by two different reaction paths having different activation energies. The reaction is said to be **kinetically controlled** if conditions are such that the product that is formed more rapidly (by the path of lower activation energy) is the dominant product. It is important that this may or may not be the more stable product.

If the reaction is carried out under conditions in which the two products can be in equilibrium with each other (**thermodynamic** or **equilibrium control**), the more stable product will always dominate. Thermodynamic control is always favored by higher temperatures and longer reaction times. See Figure E29.1.

EXERCISE

1. In energy diagram #1, R represents the reactants and P_1 and P_2 the products. Which product would dominate under kinetic control? Under thermodynamic control? Consider the same questions for energy diagram #2.

The concepts of kinetic and thermodynamic control can also be applied to the competition of two related reactants with a single reagent to form two related products, as in the following experiment. We will use the reaction of an aldehyde or ketone with semicarbazide to form a semicarbazone.

$$RCR' + H_2NNHCNH_2 \longrightarrow RR'C = NNHCNH_2 + HOH$$

Semicarbazide   A semicarbazone

More specifically, we will allow furfural and cyclohexanone to compete for a limited amount of semicarbazide under conditions that allow us to compare the relative rates of formation and the stabilities of the two semicarbazones.

Cyclohexanone   Furfural

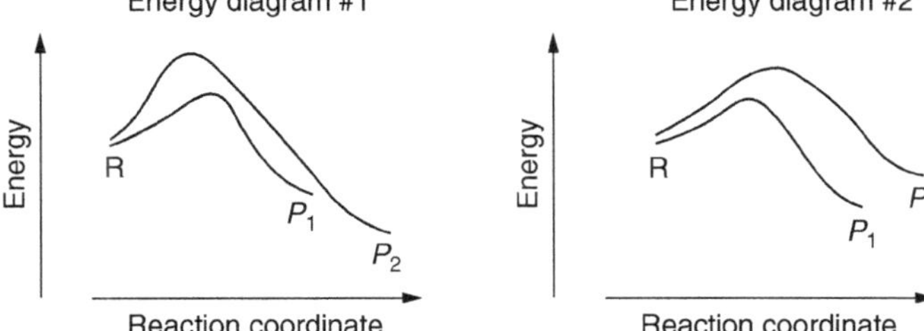

Figure E29.1 Diagram of the reaction coordinate versus the energy for a reactant going to two products

---

WASTE DISPOSAL NOTE

Waste methanol should be collected in a labeled container.

NOTE

The Instructor should distill furfural (about 5 g/student) in advance of the laboratory. The commercial aldehyde is normally contaminated with significant amounts of the carboxylic acid from air oxidation.

A. Semicarbazone Formation

Slowly dissolve 1.7 g (15.2 mmol) of semicarbazide hydrochloride in a solution of 3.0 g of sodium bicarbonate in 35 mL of water in an Erlenmeyer flask and when effervescence ceases, add 1.6 mL (15.5 mmol) of cyclohexanone. Shake or swirl the mixture for 8 min and isolate the product by suction filtration (use a suitably sized Büchner or Hirsch funnel). Recrystallize it from a minimum amount of hot water, allow it to dry, and obtain a mp.

Carry out the same procedure with 1.3 mL (15.7 mmol) of distilled furfural and a reaction time of 20–25 min with a sand bath at about 100°C. During the 20-min reaction time, begin the next experiment. Recrystallize the product from hot methanol (20 mL/g of product), allow it to dry, and obtain a mp.

B. Competitive Reactions

Using the method described in Part **A**, prepare the semicarbazone of a mixture of 0.8 mL of furfural, 1 mL of cyclohexanone, and 1 g of semicarbazide hydrochloride dissolved in a solution of 2.0 g of sodium bicarbonate in 25 mL of water. Isolate the product after 5 min, recrystallize it from a minimum amount of hot methanol (about 10 mL/g of product), and identify the product by mp and mixed mp or by comparison of its IR spectrum with that of authentic samples of the semicarbazones of cyclohexanone and furfural.

Repeat the experiment, but heat the reaction mixture (sand bath) at about 100°C for 1.5 h.

C. Interconversion Attempts

Boil a mixture of 0.5 g of cyclohexanone semicarbazone and 0.4 mL of furfural in
10 mL of water for approximately 20 min, allow the solution to cool, and collect the
precipitated semicarbazone. Recrystallize it from a minimum amount of hot methanol
(about 10 mL/g of product) and identify it.

Repeat the experiment with 0.5 g of furfural semicarbazone and 0.4 mL of
cyclohexanone.

2. Describe the results of all these experiments in terms of the concepts of
   thermodynamic and kinetic control. Assume that the energies of cyclohexanone
   and furfural are about the same and draw an energy profile for both reactions
   that would explain the behavior you observed.

Kinetic Investigation of Unimolecular Solvolysis[1]

The speed at which a reaction takes place is of particular importance in chemistry because it provides information about the detailed path over which reactants travel on their way to becoming products (the reaction mechanism). The correct interpretation of rate data depends on a knowledge of factors that can influence the rate; these include (1) the structure of the compound or compounds entering into the reaction, (2) the temperature, (3) the type of solvent in which the reaction is carried out, and (4) for many reactions, the concentrations of the reacting species.

We shall investigate some of these effects on a simple but very important process in organic chemistry, **unimolecular solvolysis**, sometimes designated S_N1. Such a reaction can be written in two steps. These steps are illustrated with the reaction we are to study in this experiment—namely, the hydrolysis of *tert*-butyl chloride (I).

Background Reading
Discussion on reaction rates or kinetics and on the S_N1 reaction in any modern organic textbook

Timing
(5 h)

$$CH_3 \underset{\underset{CH_3}{|}}{\overset{\overset{CH_3}{|}}{C}} Cl \xrightarrow{\text{Slow}} CH_3 \underset{\underset{CH_3}{|}}{\overset{\overset{CH_3}{|}}{C^+}} + Cl^- \tag{1}$$

$$\text{(I)} \qquad\qquad \text{(II)}$$

$$2HOH + \text{(II)} \xrightarrow{\text{Fast}} CH_3 \underset{\underset{CH_3}{|}}{\overset{\overset{CH_3}{|}}{C}} OH + H_3O^+ \tag{2}$$

$$\text{(III)}$$

Equation 1 represents the breaking of the carbon–halogen bond of *tert*-butyl chloride in such a way as to form a reactive intermediate carbocation (II) and a chloride ion. The carbocation then reacts rapidly with water to form *tert*-butyl alcohol (III) and a hydronium ion.

If a reaction goes through several steps, the rate of the slowest step represents the overall rate of the reaction. Thus, Equation 1 represents the rate-determining step of this reaction.

---

[1] J. A. Landgrebe, *J. Chem. Educ.*, **41**, 567 (1964).

In the experiment, a solution of *tert*-butyl chloride in acetone is quickly added to aqueous sodium hydroxide that contains a bromphenol blue indicator (blue at pH ≥ 4.6). The amount of hydroxide is equivalent to 10% of the total amount of *tert*-butyl chloride. Solvolysis begins at the time of mixing and proceeds to give *tert*-butyl alcohol and hydronium ion at a characteristic rate. When the hydronium ion has neutralized the sodium hydroxide, the indicator changes from blue to yellow and the time required for this change represents the time for 10% reaction.

A. General Procedure

Repeat the following procedure three times and record the time for 10% reaction, the solvent composition (by volume), and the temperature. Determine the average time and the average deviation in seconds.

1. Pipet 3 mL (5-mL graduated pipet) of a 0.1 *M* solution of *tert*-butyl chloride in acetone into a 25-mL Erlenmeyer flask and place the flask on white paper.
2. Pipet 0.3 mL (1-mL graduated pipet) of the 0.1 *M* sodium hydroxide solution into a second 25-mL flask and then add 6.7 mL of distilled or deionized water (small graduated cylinder or pipet). Add several drops of bromphenol blue indicator solution.
3. Note the time and quickly pour the acetone solution into the water solution, swirl for a second, and immediately pour the solution back into the other flask. (This procedure will insure complete mixing.) Note the time when the indicator changes color. *Before beginning each experiment, the two flasks should be rinsed with acetone and drained.*

EXERCISES

1. How many millimol (mmol) each of *tert*-butyl chloride and hydroxide ion are present in each flask prior to mixing?
2. Does the acetone participate directly in the reaction? Does the water participate in the reaction? Explain.
3. The relationship between the rate constant for a unimolecular reaction and the percent reaction is given by the expression

$$kt = \ln \frac{1}{1 - \%\ reaction/100}$$

Calculate the value of k (in sec^{-1}) for the data from Part **A**.

B. Concentration Effect in a Unimolecular Process

Use the experimental procedure described in Part **A**, but before the two solutions are mixed, add 10 mL of a 70% water–30% acetone (by volume) solution to the flask that contains the aqueous hydroxide. (The experiment is being carried out at half the concentration used in Part **A** but with the same overall solvent composition.) Try the reaction several times.

4. Is your result consistent with the fact that the slow step of the reaction has been classified as unimolecular? Explain. What would be the expected qualitative concentration dependence on a rate-determining step that was bimolecular?

C. Temperature Dependence

The tendency of the carbon–chlorine bond to rupture will depend on whether a sufficient amount of energy (in the stretching vibration of the bond) is available in the *tert*-butyl chloride molecules. For each of the following experiments, use the procedure of Part **A**.

1. Prepare a water bath from a large beaker and adjust the temperature (with crushed ice) to about 10°C below room temperature. Again follow the procedure of Part **A**, but before you mix the two solutions, allow the temperature of the Erlenmeyer flasks to equilibrate in the water bath for about 5 min. (A piece of copper wire wound around the neck of the flasks and hooked over the edge of the beaker provides a convenient support.) Try the reaction twice.
2. Warm a large beaker of water to about 10°C above room temperature and repeat the preceding experiment twice. Record all the results in your notebook.

EXERCISES

5. What conclusion do you draw about the effect of temperature on the S_N1 reaction rate constant? Do you think your results would be qualitatively true for other reactions?
6. Calculate the rate constant (in $\sec^{-1}$) for each of the temperatures used.

D. Solvent Effects

Using two Erlenmeyer flasks, pipet 2 mL of the 0.1 *M* solution of *tert*-butyl chloride in acetone into one of the flasks and 0.2 mL of the 0.1 *M* sodium hydroxide solution plus 7.8 mL of water and several drops of bromphenol blue indicator into the other (80% water–20% acetone). Carry out the reaction as described in Part **A** and carefully record the results.

EXERCISES

7. Which is the more polar solvent, acetone or water?
8. Note that the rate-determining step of this reaction involves the production of a positively charged particle. What effect would you expect a more polar solvent medium to have on the solvolysis rate? Explain. How does this expectation compare with the result of the experiment?
9. If time permits, design and carry out an experiment to measure the time for 10% reaction of *tert*-butyl chloride in a 60% water – 40% acetone solvent.

E. Structural Effects

Prepare appropriate solutions and carry out the procedure of Part **A** with *tert*-butyl bromide and with isopropyl or *sec*-butyl bromide. (Be certain to use *dry* acetone for preparation of the solution of the alkyl halide.) Do not wait more than 5 min for a reaction.

10. Summarize how the leaving group and the nature of the alkyl group influence the rate constant for unimolecular solvolysis.

F. Activation Energy (Optional)

Because molecules of reactant must have a certain minimum energy before a reaction can take place, we can define an energy of activation that is related to the rate constant by Equation 3, in which c is a constant, E_a is the activation energy in calories per mol, R is the gas constant (1.99 cal/K mol), and T is the absolute temperature.

$$k_1 = ce^{-E_a/RT} \tag{3}$$

By taking the natural logarithm of both sides of Equation 3, we obtain Equation 4, which indicates that a plot of ln k versus $1/T$ should give a straight line with slope equal to $-E_a/1.99$. (Note that Equation 4 has the form $y = mx + b$, the general equation for a straight line in which m is the slope, defined as $\delta y/\delta x$.)

$$\ln k_1 = -(E_a/1.99)(1/T) + \ln c \tag{4}$$

Plot the data from Exercises 3 and 6 and record your approximate value for E_a in kcal/mol. A typical plot is illustrated in Figure E15.1; note the negative slope.

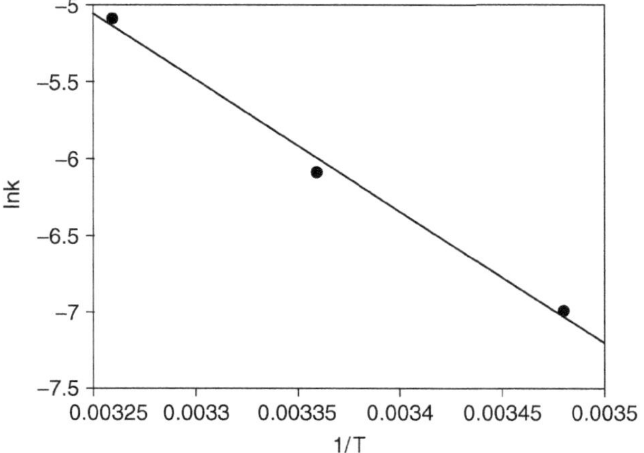

Figure E15.1　An Arrhenius plot

11. The expression for the rate of disappearance of alkyl chloride is given by

$$\frac{-d[\text{RCl}]}{dt} = k_1[\text{RCl}]$$

Carry out the following integration, in which $[\text{RCl}]_0$ is the initial concentration of alkyl chloride and $[\text{RCl}]$ is the concentration of alkyl chloride at time $= t$.

$$-\int_{(\text{RCl})_0}^{(\text{RCl})} \frac{d[\text{RCl}]}{[\text{RCl}]} = \int_0^t k_1\, dt$$

We can see from the definitions of $[\text{RCl}]_0$ and $[\text{RCl}]$ that the fraction of reaction $= ([\text{RCl}]_0 - [\text{RCl}])/[\text{RCl}]_0 = \%$ reaction/100. This simplifies to the form

$$\frac{[\text{RCl}]}{[\text{RCl}]_0} = 1 - \frac{\% \text{ reaction}}{100}$$

From this relationship and the results of your integration, derive the equation in Exercise 3.

Fischer Esterification

Background Reading

Be certain to read all of Parts **A** and **B** below before you begin. A general discussion of ester formation from carboxylic acids and alcohols should be consulted in any organic textbook

Timing

(4 h)

The reaction of an alcohol and a carboxylic acid to form an ester and water is known as the Fischer esterification and can be described in equilibrium terms as shown.

$$RCOOH + ROH \overset{H^+}{\rightleftharpoons} RCOOR + HOH$$

$$K = \frac{(\text{Ester})(\text{Water})}{(\text{Carboxylic acid})(\text{Alcohol})}$$

Carried out in the presence of a strong acid catalyst, the reaction can be driven toward completion with an excess of one of the reactants or removal of water or the ester as it forms.

EXERCISES

1. Under what circumstances would it be impractical to use an excess of ROH to drive the equilibrium toward products? What alternatives for the synthesis of esters are available?

2. Why is it necessary to include water in the equilibrium expression instead of including it as part of the value of K as is done with most equilibrium expressions of weak acids in water?

CAUTION!

Concentrated sulfuric acid is highly corrosive. Plastic gloves are recommended.

NOTE

Read the following paragraphs, carefully outline your work-up procedure, and check this information with your instructor before you continue.

A. Equilibration

Start with about 25 mmol of benzoic acid (weighed to the nearest mg), a fivefold molar excess of methanol, and about 0.5 mL of concentrated sulfuric acid (added cautiously to the methanol). Use a 25-mL pear-shaped flask (slightly larger will work) with a boiling chip and a condenser, and carry out the esterification for 1 h at reflux (Figure E43.1).

Cool the reaction to about room temperature with an ice bath and carefully dilute the reaction mixture with just enough water until a phase separation occurs. Add 5 mL of methylene chloride, transfer the mixture to a small separatory funnel, shake it, and separate the lower organic layer. Extract the aqueous layer (which contains most of the sulfuric acid catalyst and some of the methanol), separate the layer, and extract it with one 10-mL and one 5-mL portion of methylene chloride. Combine all of the methylene chloride extracts and extract them with two 4-mL portions of saturated sodium bicarbonate solution. Add anhydrous magnesium sulfate to the methylene chloride solution, swirl it, stopper the container, and set it aside to dry.

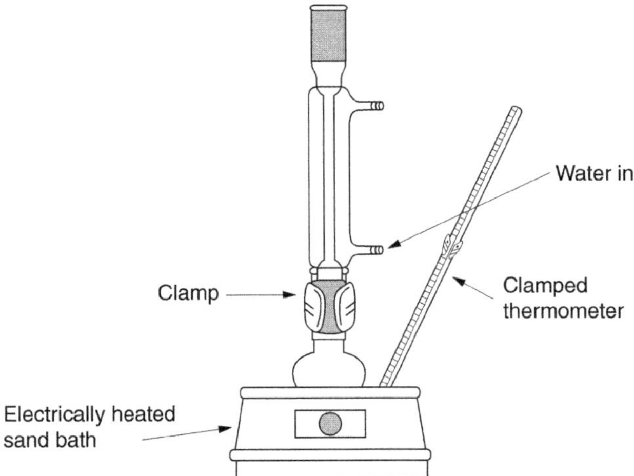

Figure E43.1 Apparatus for reflux in Part **A**

Combine the bicarbonate extracts in a small beaker. Precipitate the unreacted benzoic acid by adding the smallest possible amount of concentrated hydrochloric acid (use pH paper to test for acidity), cool the slurry in an ice bath, and isolate the benzoic acid by suction filtration on a Hirsch funnel. A very small amount of ice water can be used to help transfer the benzoic acid to the funnel. Suck the solid dry for a few minutes, transfer it to a tared piece of glazed paper, dry the sample with gentle heat (heat lamp from above), and obtain the weight by difference.

The dried methylene chloride solution of methyl benzoate should be filtered and evaporated (rotary evaporator or the method shown in Figure E43.2) to give the crude liquid ester.[1] The ester should be vacuum distilled (aspirator or diaphragm pump) with an apparatus such as that shown in Figure E43.3. If you know the approximate pressure (from a manometer),[2] use the nomograph in Figure 3.10 to estimate the boiling point of methyl benzoate at the reduced pressure. If time permits, check the purity of your ester product by GC.

1. Although a 50-mL flask can be used, it might be better to evaporate the solution in a 25-mL flask by adding it in several batches. This eliminates the need to transfer the crude product to a smaller flask for the distillation.

2. If you are using an aspirator, you can measure the water temperature and assume that the reduced pressure is the same as the vapor pressure of water at that temperature. Tables for the vapor pressure of water at various temperatures can be found in the *Handbook of Chemistry and Physics*, published by the Chemical Rubber Company.

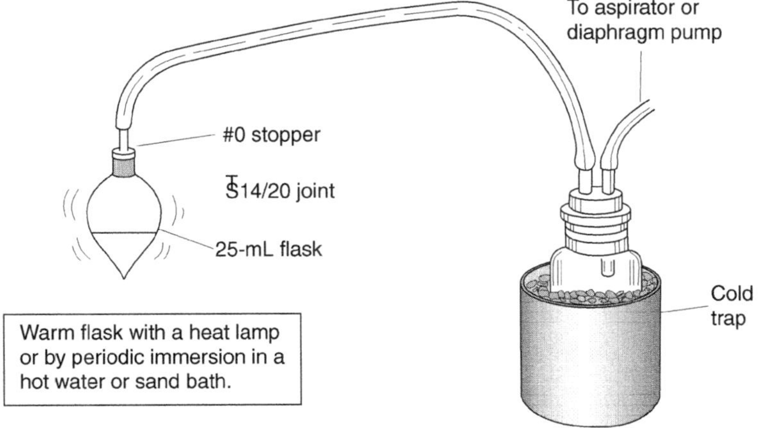

Figure E43.2 Simplified solvent evaporation

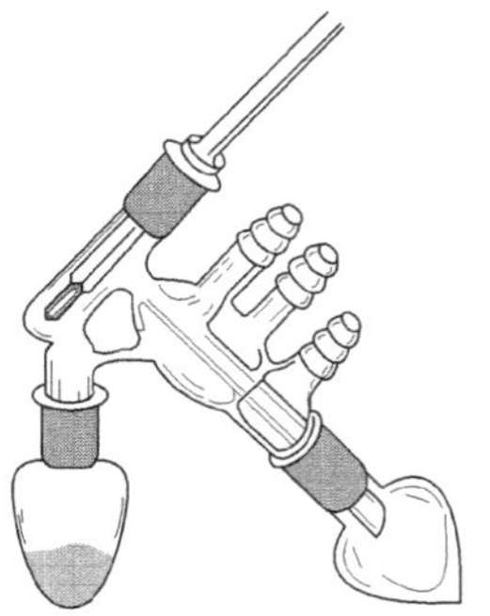

Figure E43.3 Apparatus for vacuum distillation of product ester (use a sand bath)

B. Equilibrium Constant Calculation

Correct the amount of recovered benzoic acid by estimating the amount of benzoic acid dissolved in the solution from which it was precipitated. Use the volume of aqueous solution from which the acid was precipitated and the approximate solubility of benzoic acid in water at room temperature of 3.4 g/L. Then from the resulting total weight of benzoic acid (recovered and dissolved) at equilibrium, (PhCOOH), the initial amount of benzoic acid, $(PhCOOH)_0$, and the initial amount of methanol, $(MeOH)_0$, the approximate equilibrium constant for this reaction can be calculated. First define $(PhCOOH)_0 - (PhCOOH) = P$, the amount of either product, methyl benzoate or water at equilibrium. The expression for the equilibrium constant then becomes:

$$K_{eq} = \frac{P^2}{(PhCOOH)[(MeOH)_0 - P]}$$

C. Infrared Spectrum

Obtain an IR spectrum on the methyl benzoate product to turn in with your sample. Clearly indicate peaks that provide evidence for the presence of an aromatic ring.

9-Benzal- and 9-Benzylfluorene; Aldol and Cannizzaro-Type Reactions[1]

Background Reading
Material on the aldol condensation and the Cannizzaro reaction will provide information about reactions similar to those carried out in this experiment.

Timing
(5 h)

The reaction of a carbanion with a carbonyl group provides a useful method for the formation of new carbon–carbon bonds. In this experiment, a condensation between

$$R^- \; + \; \underset{X \quad Y}{\overset{\overset{\textstyle O}{\|}}{C}} \quad \longrightarrow \quad R-\underset{X}{\overset{\overset{\textstyle O^-}{|}}{C}}-Y$$

fluorene and benzaldehyde will be done and this will be followed by reduction of the resulting carbon–carbon double bond of the 9-benzalfluorene product by means of a hydride transfer reaction. The reactions will be carried out by two different procedures, one that requires isolation of the intermediate 9-benzalfluorene and one that does not.

Fluorene + PhCHO $\xrightarrow[\substack{PhCH_2OH \\ 100°C}]{KOH}$ H$_2$O + 9-Benzalfluorene (C=CHPh)

---

1. Adapted from I. Angres and H. E. Zieger, *J. Chem. Educ.*, **51**, 64 (1974).

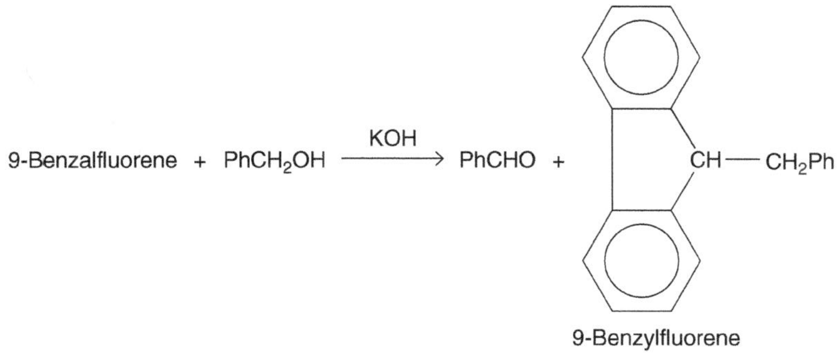

9-Benzalfluorene + PhCH$_2$OH $\xrightarrow{\text{KOH}}$ PhCHO + [9-benzylfluorene structure] CH—CH$_2$Ph

9-Benzylfluorene

The first step in the procedure in which 9-benzalfluorene will be isolated involves equilibrium formation of the 9-fluorene anion followed by attack by the anion on the carbonyl group of benzaldehyde. Elimination of a molecule of water then leads directly to 9-benzalfluorene, which is isolated and purified (Part **A**).

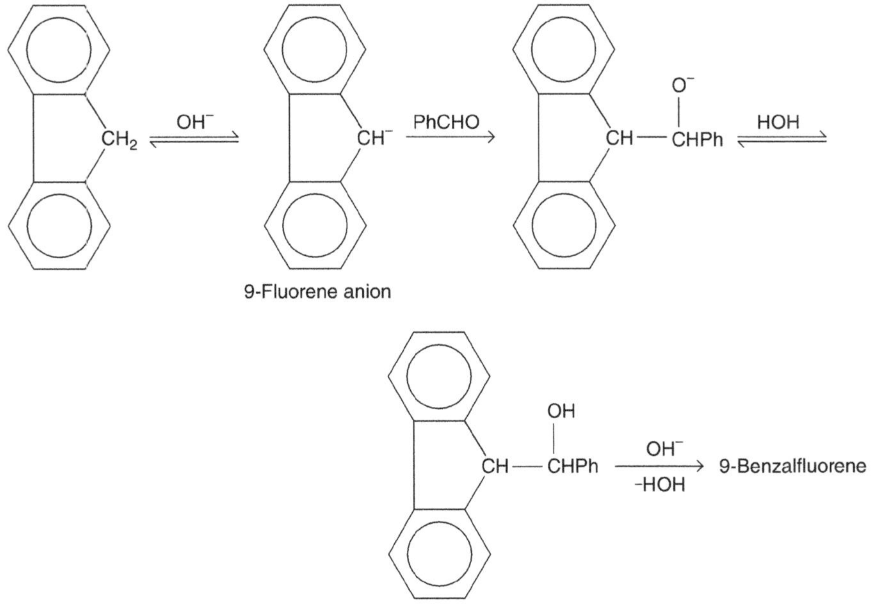

[structure] CH$_2$ $\xrightleftharpoons{\text{OH}^-}$ [structure] CH$^-$ $\xrightarrow{\text{PhCHO}}$ [structure] CH—CHPh (O$^-$) $\xrightleftharpoons{\text{HOH}}$

9-Fluorene anion

[structure] CH—CHPh (OH) $\xrightarrow[\text{-HOH}]{\text{OH}^-}$ 9-Benzalfluorene

The purified compound provides the starting material for reduction of the **exocyclic** (attached but external to a ring) carbon–carbon double bond by transfer of hydride ion from the anion of benzyl alcohol to the 9-benzalfluorene (Part **C**).

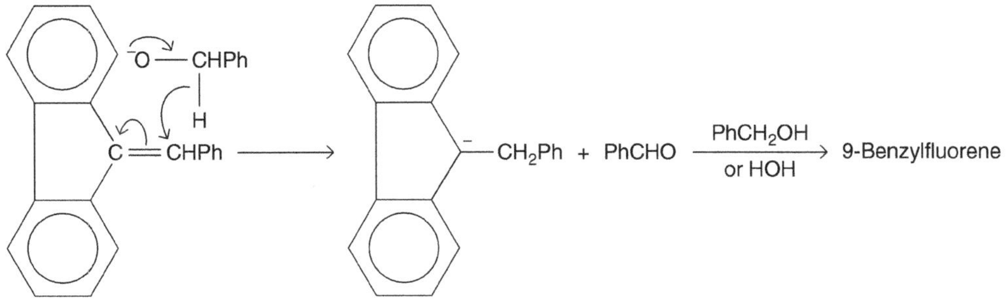

[structure] $^-$O—CHPh, H, C=CHPh $\longrightarrow$ [structure] —CH$_2$Ph + PhCHO $\xrightarrow[\text{or HOH}]{\text{PhCH}_2\text{OH}}$ 9-Benzylfluorene

If, instead of treating fluorene with a molar equivalent of benzaldehyde and an excess of strong base in benzyl alcohol at 100°C (Part **A**), it is treated with a trace of benzaldehyde and an excess of strong base in benzyl alcohol at about 200°C (Part **B**), the result is a good yield of the reduced product, 9-benzylfluorene. Under the latter conditions any 9-benzalfluorene formed is immediately reduced to 9-benzylfluorene and a molar equivalent of benzaldehyde, then condenses with fluorene to produce more 9-benzalfluorene. This cyclic sequence of reactions continues until all the fluorene has been consumed.

$$\text{Fluorene} + \text{PhCH}_2\text{OH} \xrightarrow[\text{trace PhCHO}]{\text{KOH, 200°C}} \text{9-benzylfluorene} + \text{H}_2\text{O}$$

1. Account for the ease of formation of a 9-fluorene anion relative to a benzylic anion, PhCH_2^-, which does not form as easily.
2. At what other position on the 9-benzalfluorene might the hydride attack? Design an experiment to prove which position(s) on the 9-benzalfluorene is (are) being attacked.

NOTE

When it is in contact with atmospheric oxygen, benzaldehyde is slowly oxidized and is invariably contaminated with small amounts of benzoic acid. Although the best yields of 9-benzalfluorene are obtained with freshly distilled benzaldehyde, the procedure described works quite well with commercial benzaldehyde that has been kept in tightly capped bottles.

CAUTION!

Solid potassium hydroxide is quite corrosive. Handle the pellets with a spatula or tweezers; wear plastic gloves.

A. 9-Benzalfluorene

Into a 50-mL or 100-mL round-bottomed flask, place 4.15 g (25 mmol) of fluorene, 4 mL (4.16 g, 39.2 mmol) of benzaldehyde, 20 mL of benzyl alcohol, and 2.0 g (35.7 mmol) of potassium hydroxide. Attach a condenser and maintain the reaction mixture at about 100°C in a sand bath for 90 min (Figure E48.1).

NOTES

1. Potassium hydroxide is quite hygroscopic. It should be weighed quickly and transferred immediately into the flask. Be certain to recap the original container without delay.
2. During the heating time for Part **A**, prepare the reaction mixture for Part **B**. Begin the reflux for Part **B** as soon as the condenser and sand bath from Part **A** are available.

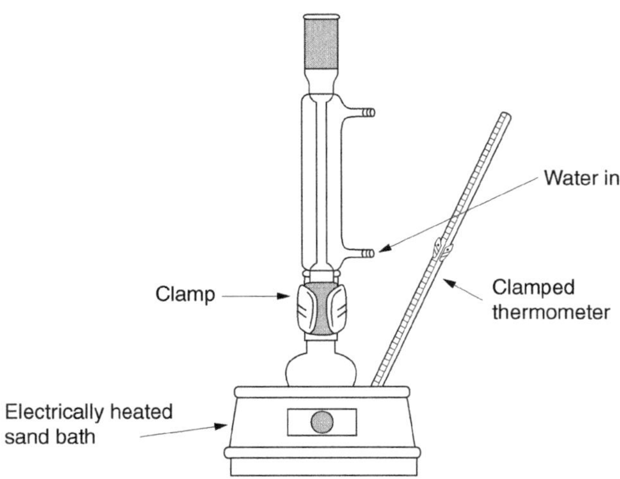

Figure E48.1 Apparatus for reflux

When the 90-min heating period is over, allow the reaction mixture to cool, transfer it to a 500-mL Erlenmeyer flask, and add 200 mL of water. (Some of the water should be used to rinse the reaction vessel.) The product crystallizes slowly over a period of 30 min.

The crude yellow crystalline product should be suction filtered and washed with plenty of water. Leave the product in the Büchner funnel for a few min (suction on) and then spread it on glazed paper to dry, mp approximately 68–74°C.

During the time you are waiting for the 9-benzalfluorene to dry, recrystallize your sample of 9-benzylfluorene as described in Part **B**.

The dry 9-benzalfluorene can be recrystallized from hot heptane by using about 4 mL of heptane per g of crude product. Cool the heptane in ice before isolating the crystallized product by suction filtration. Wash the collected product with only a very small amount of cold heptane. Pure pale-yellow 9-benzalfluorene has a mp of 75–76°C. The yield will be in the range 30–70%.

> **NOTE**
>
> If globules of oil form rather than crystals, carefully decant (and discard) the upper water layer and add 5–10 mL of 95% ethanol. Crystallization will then occur.

B. 9-Benzylfluorene

Into a 50- or 100-mL round-bottomed flask, place 4.15 g (25 mmol) of fluorene, 10 mL of benzyl alcohol, 2–3 drops of benzaldehyde, and 2 g of potassium hydroxide. Attach a condenser and maintain the reaction mixture at reflux for 1 h (Figure E48.1). Allow the reaction mixture to cool, add 10 mL of water, and isolate the crude 9-benzylfluorene by suction filtration. Wash the crystals with plenty of water, allow them to remain in the Büchner funnel for a few minutes, and then spread them on a piece of weighing paper to dry. Set the crystals aside while you carry out other parts of the experiment.

The dry crude product will have a mp of 120–130°C. The crude product should be recrystallized from hot heptane by using about 18 mL of heptane per g of compound to give a white crystalline product with mp 132–34°C. Cool the heptane solution in ice before isolating the crystallized product. The yield should be in the range 50–90%.

C. Reduction of 9-Benzalfluorene

Into a 25- or 50-mL round-bottomed flask place 2.0 g (7.9 mmol) of the purified 9-benzalfluorene (Part **A**), 6 mL of benzyl alcohol, and 0.5 g (8.9 mmol) of potassium hydroxide. Attach a condenser and heat the reaction mixture at reflux for 15 min (Figure E48.1). A precipitate of potassium benzoate will form. Allow the reaction mixture to cool, add 20 mL of water, and swirl the flask vigorously. A white crystalline product should form. In the event that only an oil results, carefully decant and discard the upper aqueous layer and add 2–3 mL of 95% ethanol to the oil. Swirl the resulting mixture and crystallization will take place. Isolate the crystalline 9-benzylfluorene by suction filtration, wash the product with water, allow it to suck dry as much as possible, and then spread it on glazed paper to dry. Obtain the mp and turn your sample in to your instructor without further purification.

EXERCISES

3. Compare the overall yield of 9-benzylfluorene from fluorene by the two procedures used in these experiments. Comment on possible reasons for the difference.
4. What is the origin of the potassium benzoate that forms in Part **C**?
5. **(Optional; Spectroscopic)** How would you use either ^{1}H NMR or ^{13}C NMR to distinguish between 9-benzalfluorene and 9-benzylfluorene?

Oxidation of (–)-Borneol to (–)-Camphor; Specific Rotation

Background Reading
Read about various ways to oxidize alcohols to ketones in any organic textbook.

Timing
(5 h)

The following experiment illustrates the oxidation of a secondary alcohol to a ketone and provides an opportunity to use thin-layer chromatography, infrared spectroscopy, and optical rotation measurements to derive information about the structure and purity of an isolated product. Additional techniques used include extraction and sublimation.

The unbalanced equation for the oxidation process is shown below.

(–)-Borneol   (–)-Camphor

CAUTION!

Glacial acetic acid is corrosive. Plastic gloves are recommended.

1. Write a balanced equation for the oxidation of (–)-borneol to (–)-camphor with HOCl (hypochlorite in acid) going to HCl.

A. Preparation of (–)-Camphor

Dissolve 500 mg (3.25 mmol) of (–)-borneol in 1.5 mL of glacial (concentrated) acetic acid in a 25-mL Erlenmeyer flask with swirling at room temperature. Add about 5.0 mL of fresh household bleach solution (for example, clorox^{TM}, which is about 5.25% NaOCl) in one batch with swirling. The solution will become noticeably warm and product will precipitate. Swirl the reaction mixture occasionally over the next 15 min and test the sample after 10 min by placing one drop on wet starch-iodide paper. A blue color indicates excess hypochlorite. If the test is negative, add a small amount of additional hypochlorite solution, swirl, let stand 5 min, and test it again.

$$I^- \xrightarrow{\text{oxidation}} I_2 \xrightarrow{\text{starch}} \text{blue starch-iodine complex}$$

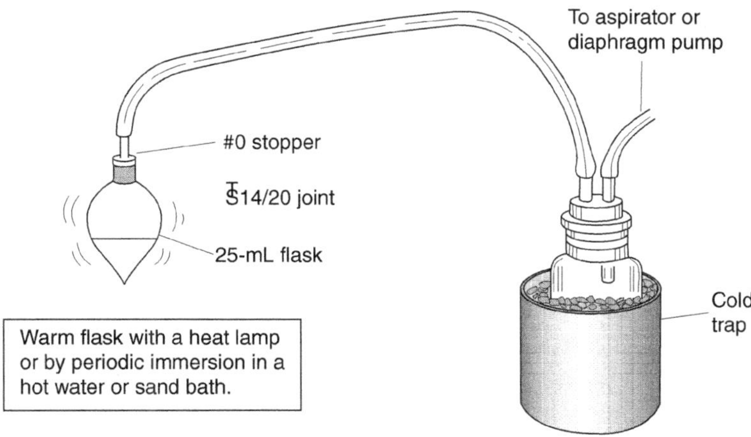

Figure E26.1 Simplified method for solvent evaporation

Dilute the reaction mixture with about 15 mL of water and transfer it to a 60-mL separatory funnel with 15 mL of diethyl ether. Add 0.5–1.0 mL of saturated sodium bisulfite solution to reduce any residual hypochlorite (negative test with starch-iodide paper) and shake the mixture for at least 20–25 sec before draining the lower layer. Wash the ether layer with about 8 mL of saturated sodium bicarbonate solution (gas evolution).[1] Drain off the lower aqueous layer, and transfer the ether to a small Erlenmeyer flask (use a little ether rinse). Dry the ether with a small amount of anhydrous Na_2SO_4 (allow at least 10 min), decant the solution into a tared, 25-mL Erlenmeyer flask connected to a diaphragm pump or aspirator (Figure E26.1), and carefully evaporate the ether. Use a heat lamp to warm the flask, *but only to room temperature*. Weigh the flask and calculate the crude yield of camphor; it should be very good.

$$HOCl + HSO_3^- \longrightarrow Cl^- + 2\,H^+ + SO_4^-$$

Save a portion (1–2 mg) of this product for a melting point and a second portion (about 10 mg) for determination of the IR spectrum. Weigh and sublime the remaining portion of the product in an apparatus such as that pictured in Figure E26.2, but *without* vacuum. The central test tube in the apparatus must be kept cold with ice water, but condensation of moisture from the atmosphere should be avoided. Heat the camphor carefully with a warm sand bath. Remove the central cold finger of the sublimator carefully so as not to detach any camphor and transfer it to a tared vial with a spatula. Determine the yield of the sublimed product. Take the mp of the sublimed product in a *sealed* capillary. Pure, dry (–) or (+)-camphor completely melts at 177–178°C.

1. Check the pH of the aqueous layer to make certain it is basic. If it is neutral or acidic, do a second rinse with a few mL of bicarbonate solution.

> **IMPORTANT NOTE**
>
> Do not heat the flask or apply vacuum after the solvent is gone. Camphor has a high vapor pressure and will sublime easily.

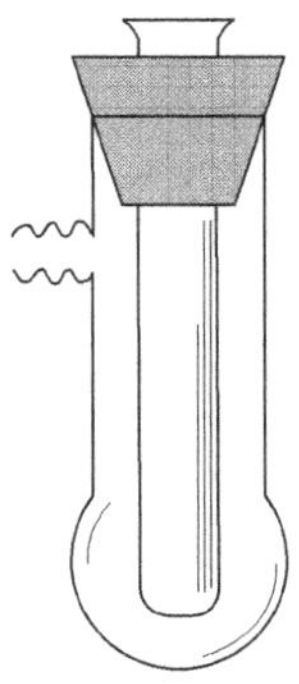

Figure E26.2 Apparatus for sublimation

2. Borneol can also be oxidized to camphor with other oxidizing agents, such as sodium dichromate in acid. Write a balanced equation for this oxidation ($Cr_2O_7^{2-}$ is reduced to Cr^{3+}). What weight of $Na_2Cr_2O_7 \cdot 2\,H_2O$ should be needed (theoretically) to oxidize 0.50 g of borneol?

B. Infrared Spectrum

Take the IR spectra of the crude and purified products as a thin film on Teflon tape.

EXERCISES

3. What conclusions can be drawn about the presence or absence of unreacted starting material?
4. Which compound, (–)-borneol or (–)-camphor, has the highest vapor pressure? (*Hint: What are the listed boiling points?*)
5. Why does the carbonyl stretching frequency in the IR spectrum of camphor occur near 1740 cm^{-1}, whereas that of acetophenone ($C_6H_5COCH_3$) is found at 1680 cm^{-1} and that for cyclohexanone is found at 1710 cm^{-1}?

C. Thin-Layer Chromatography

Run a TLC comparison of samples of authentic borneol, authentic camphor, and your purified camphor with methylene chloride (reagent grade) as the developing solvent. The samples should be prepared as concentrated solutions in small amounts (100 μL) of methylene chloride in three labeled vials.

EXERCISE

6. (**Optional**) Careful observation of the TLC plate may reveal a small spot with a slightly longer R_f value than that for camphor. The structure of this impurity is not known. Suggest a method for isolating and purifying this compound so that spectroscopic techniques could be used to determine the structure.

WASTE DISPOSAL NOTE

The ethanol solution from the rotation measurement can be flushed with excess water.

D. Optical Rotation

Weigh (to the nearest mg) 125–150 mg of purified camphor into a tared 5-mL volumetric flask and fill it to the mark with ethanol. Be certain to mix the solution by inverting the flask with the stopper in place. Fill a 1-dm (center-fill) polarimeter tube and use the automatic polarimeter to obtain a measured rotation at the sodium-D line (589 nm). Calculate the specific rotation and compare it to the literature value for optically pure (–)-camphor of –43.8 in ethanol.[2] Look up the specific rotation of (–)-borneol in ethanol and comment on whether or not a small amount of the unreacted starting material would affect your rotation measurement.

2. Small amounts of water in the ethanol will result in a slightly lower value.

4-Methylbenzophenone; Friedel–Crafts Acylation[1]

The reaction between a carboxylic acid chloride and aluminum chloride results in the formation of an acyl cation salt that, because of its electron deficiency, readily attacks electron-rich aromatic rings to form the corresponding aryl ketones.

$$C_6H_5-COCl + AlCl_3 \longrightarrow C_6H_5-CO^+ AlCl_4^-$$

$$C_6H_5-CO^+ AlCl_4^- + C_6H_5-CH_3 \longrightarrow CH_3-C_6H_4-CO-C_6H_5 + HAlCl_4$$

4-Methylbenzophenone

Background Reading
Material on the Friedel–Crafts reaction in any modern text

Timing
(5 h, not counting the final crystallization, which should be allowed to proceed slowly until the next laboratory period)

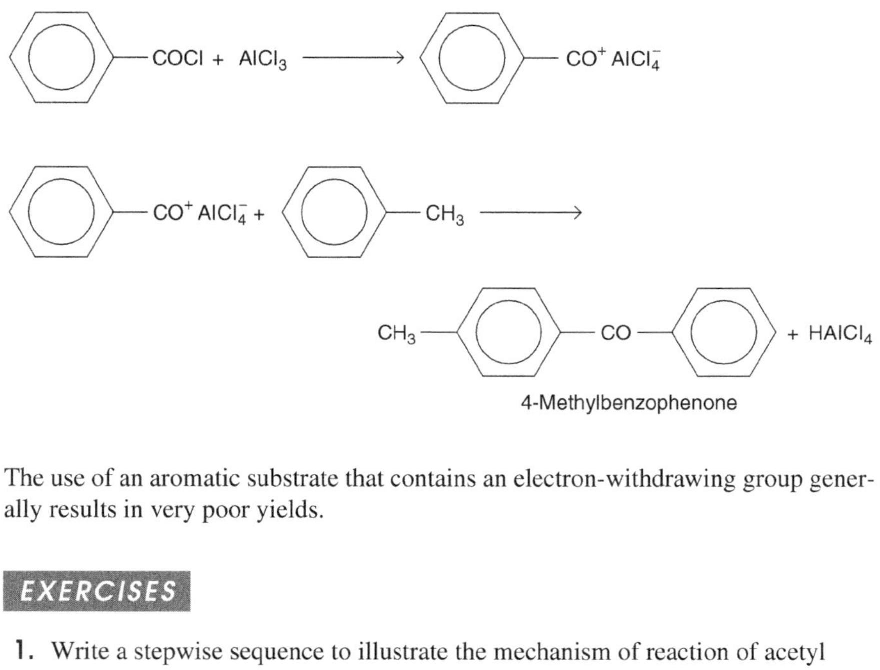

The use of an aromatic substrate that contains an electron-withdrawing group generally results in very poor yields.

EXERCISES

1. Write a stepwise sequence to illustrate the mechanism of reaction of acetyl chloride and aluminum chloride with biphenyl. Why is the major product the *para* isomer?

2. Acid anhydrides can be used in place of acid chlorides in the Friedel–Crafts reaction. Write a balanced equation for the reaction between acetic anhydride, boron trifluoride, and anisole (methyl phenyl ether).

3. What reagents are required to synthesize 3-nitrobenzophenone via the Friedel–Crafts process?

---

1. Adapted from N. M. Zaczek, J. C. Ruff, A. H. Jakewitz, and D. F. Roswell, *J. Chem. Educ.*, **48**, 257 (1971).

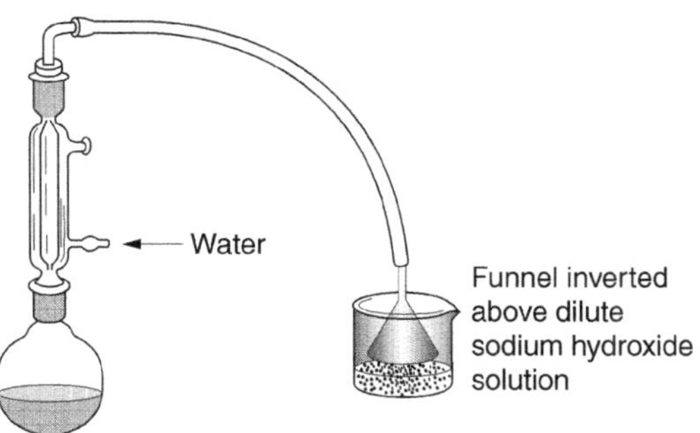

Figure E34.1 Vapor entrainment in Part **A**

A. Acylation

> **CAUTION!**
>
> Anhydrous aluminum chloride is a corrosive solid that reacts vigorously with water. Gloves are recommended.

If possible, carry out the reaction in a hood and be certain to use the simple vapor entrainment procedure shown in Figure E34.1 in order to trap hydrogen chloride produced during the reaction. Clamp the condenser loosely and allow the flask to rest on a cork ring. Charge a 100-mL round-bottomed flask with 7.0 g (0.052 mol) of anhydrous aluminum chloride and 25 mL (0.236 mol) of toluene, attach the reflux condenser, and slowly add 6 mL (0.052 mol) of benzoyl chloride in small portions through the condenser. Be certain to replace the stopper on the top of the condenser immediately after each addition. Swirl the flask after each portion has been added; hydrogen chloride will be evolved.

When the benzoyl chloride addition has been completed, clamp the flask and condenser to the ring stand or rack and maintain the solution at reflux with a sand bath for 45 min (Figure E34.2).

B. Product Isolation

> **CAUTION!**
>
> Concentrated hydrochloric acid is corrosive. Plastic gloves are recommended.

Allow the reaction mixture to cool for a few minutes and then (*in a hood*) pour the warm solution into a 250-mL Erlenmeyer flask containing 20 mL of concentrated hydrochloric acid and about 50 mL of crushed ice. Swirl the mixture, transfer it to a separatory funnel, and drain the lower, acidic water layer into a beaker or flask (save it).

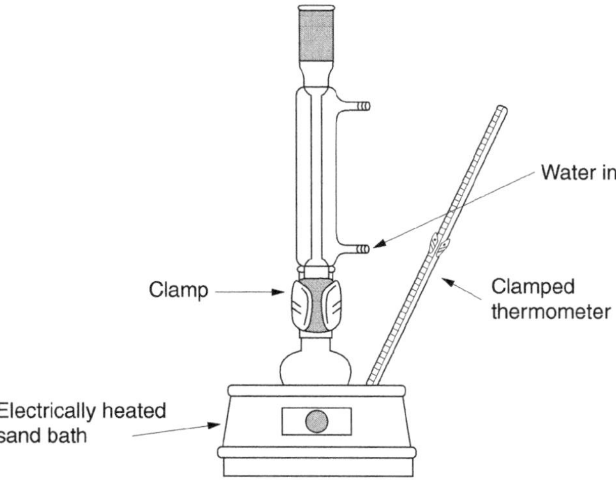

Figure E34.2 Reflux operation in Part **A**

Wash the toluene layer with a solution of 20 mL of concentrated hydrochloric acid plus 5 mL of water, combine the aqueous acidic layers, and extract them with 10 mL of toluene. Carefully wash (with swirling, not shaking, to avoid emulsion formation) the combined toluene layer with 10 mL of approximately 1.5 M (5–6%) sodium hydroxide solution and then 10 mL of sodium chloride solution.

WASTE DISPOSAL NOTE

Waste toluene collected in the cold trap of a rotary evaporator should be placed in a labeled container for nonhalogenated waste solvents.

Drain and discard the aqueous layers. Evaporate the excess toluene using a rotary evaporator attached through a cold trap to a diaphragm pump or aspirator. The flask can be heated with a hot-water bath or heat lamp. Transfer the residual liquid into a 25-mL pear-shaped flask half full of loosely packed glass wool, and carry out a simple vacuum distillation with the short-path still shown in Figure E34.3 with a sand bath as the heat source. An aspirator or diaphragm pump pressure of 30 mm would lower the boiling point of about 312–316°C at 1 atm to about 190°C, a temperature that can be easily achieved with the sand bath. Because of the high temperature and the possibility that the product could solidify, no cooling water should be used in the condenser. (Do not attempt to distill the last 0.1 mL of liquid in the flask or decomposition can occur, which will contaminate the distillate.)

Cool the distillate in ice and either scratch the inside of the vessel with a glass rod or preferably add a seed crystal of the product (obtained from your instructor). Filter the resulting slush under vacuum with a suitably sized Büchner or Hirsch funnel. This will remove oil which contains 2-methylbenzophenone.

Dissolve the solid in a minimum amount of hot 80% ethanol, filter (by gravity) with a short-stemmed funnel, and when the resulting filtrate has cooled to 25°C, add a seed crystal of the product. Allow the 4-methylbenzophenone to crystallize slowly and isolate it by suction filtration, mp 57–58°C. A second crop of crystals can be obtained by heating the final filtrate, adding water until cloudiness just persists, warming to produce a clear solution again, allowing the solution to cool, and adding a seed crystal.

Figure E34.3 Short-path still for vacuum distillation in Part **B**

EXERCISE

4. (**Optional; Spectroscopic**) Describe characteristic absorptions in the IR, ^{1}H NMR, and ^{13}C NMR spectra that would permit 4-methylbenzophenone to be detected in the crude reaction mixture.

Electronic Effects of Substituents in Reactions; Acidity Constant Determination

Background Reading
Material on resonance and inductive effects of substituents on aromatic rings in any organic textbook

Timing
(3 h)

The Hammett equation, discovered empirically, indicates a simple relationship between the logarithms of the equilibrium, or rate, constants for various reactions of substituted aromatic compounds and two constants, one characteristic of the substituent and one characteristic of the type of reaction.

$$\log\left(\frac{K}{K_0}\right) = \sigma\rho$$

In the equation, K represents the equilibrium (or rate) constant for some reaction of a specific substituted benzene derivative; K_0 represents the equilibrium (or rate) constant for the parent compound (substituent = hydrogen); ρ (rho) is the reaction constant; and σ (sigma) the substituent constant.

For the ionization of benzoic acids in water at 25°C, the value of ρ is defined as 1.00.

$$\text{Z—C}_6\text{H}_4\text{—COOH} + \text{H}_2\text{O} \underset{\rho \equiv 1.00}{\overset{K}{\rightleftharpoons}} \text{Z—C}_6\text{H}_4\text{—COO}^- + \text{H}_3\text{O}^+$$

Thus, measurement of the acidity constant for various substituted benzoic acids will lead to a direct determination of the σ constant, the value of which is proportional to the electron-donating or -withdrawing ability of the substituent group on the ring.

In the following experiments you will determine and interpret the values of sigma for several substituents.

A. Titration

Prepare dilute solutions (about 0.01 M) of benzoic acid, m- and p-hydroxybenzoic acids, and p-chlorobenzoic acid in 50% (by volume) aqueous ethanol. Place about 25 mL of each acid solution in a 100-mL beaker having a magnetic stirring bar, and titrate slowly with approximately 0.05 M sodium hydroxide in 50% (by volume) aqueous ethanol. Neither the concentration of the sodium hydroxide solution nor those for the aromatic acid solutions need to be known accurately. Follow the titration by means

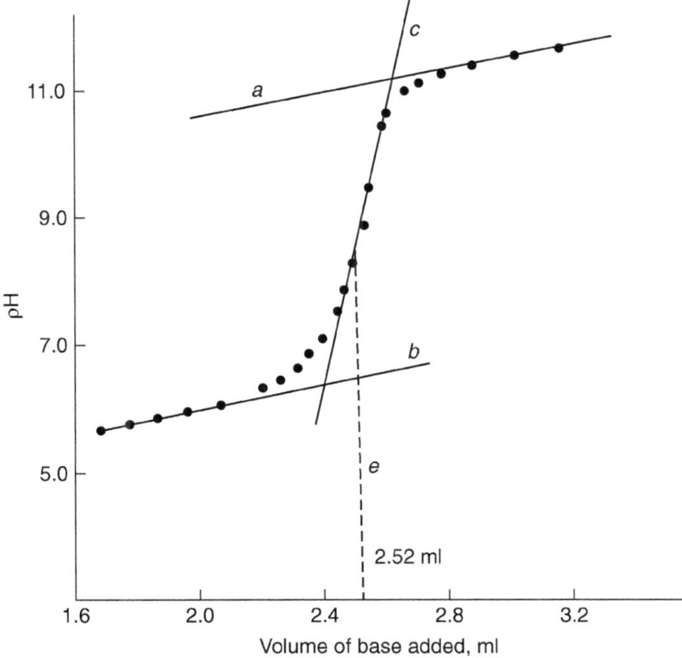

Figure E41.1 Potentiometric titration curve of a substituted benzoic acid showing a method for determination of the pK_a as described in Part **B**

of a pH meter (glass and calomel electrodes) and prepare a plot such as that shown in Figure E41.1 for each acid. Record the temperature at which the titrations were run.

EXERCISE

1. Why isn't it necessary to determine the exact concentrations of the acid or basic solutions? (*Hint: See Part* **B**.)

B. Acidity Constant Determination

If we consider the dissociation equilibrium for an acid, it becomes apparent that at the point of half-neutralization, the pH of the medium will be numerically equal to the pK of the acid.

$$\frac{(\text{ArCOO}^-)(\text{H}_3\text{O}^+)}{(\text{ArCOOH})} = K$$

$$\log\frac{(\text{ArCOO}^-)}{(\text{ArCOOH})} + \log(\text{H}_3\text{O}^+) = \log K$$

At half-neutralization $\log[(\text{ArCOO}^-)/(\text{ArCOOH})] = 0$, so that after multiplying each side of the resulting equation by -1, we obtain

$$-\log(\text{H}_3\text{O}^+) = -\log K$$

or
$$\text{pH} = \text{p}K$$

From the titration curve from each acid, construct parallel lines *a* and *b* tangential to the initial and final portions of the curve and construct line *c*. Drop a perpendicular line *e* from the midpoint of line *c* to the abscissa in order to find the volume of base needed for complete neutralization. Determine the pH corresponding to that point on the curve directly above one-half of the volume of base needed for complete neutralization. This will be the pK of the acid in this solvent system.

If difficulty is encountered in determining the exact equivalence point, an alternative method can be used in which the ordinate of the graph is the rate of change of the pH with base addition (Δ pH/Δ base) and the abscissa is the volume of base added (mL). The value of ΔpH/Δ base is easily obtained by taking the difference in pH between two data points and dividing it by the amount of base solution added between those same two points. The curve thus produced will have a maximum at the volume of added base, which corresponds to the equivalence point.

From the value $\rho = 1.46$ for substituted benzoic acid ionizations in 50% (vol.) aqueous ethanol (in contrast to the value of 1.00 for water), calculate the σ values of each of the substituents.

2. Why might you expect σ_m and σ_p values for the same substituent to be different?

3. The value of σ for hydrogen is 0.0 by definition. What is the general interpretation of values that are more negative than that for hydrogen? More positive?

4. Comment briefly on each of the σ values that you have obtained. Which substituent constants reflect mostly inductive effects? Which reflect a combination of resonance and inductive effects? What can you tell about the electronic nature of a substituent if the *meta* and *para* σ values have opposite signs? Is the value for the *para* hydroxyl group consistent with your knowledge of the electronegativity of oxygen? Explain.

5. Would you expect the value of σ for the *meta* chloro group to be more or less positive than that for the *para* chloro group? Explain.

Synthetic Dyes

Background Reading
Material on diazonium
coupling reactions, and dyes
in any organic textbook

Timing
(5 h)

Ever since the chance discovery by William Henry Perkin in 1856 that the oxidation of aniline by potassium dichromate produced a beautiful purple compound (later named "mauve"), the chemistry of synthetic dyes has been developed extensively and has become an important facet of modern technology. Although the structures of different dyes differ considerably, they are all characterized by an extended system of conjugated multiple bonds. An example is provided by methyl orange, an **azo dye** that exists in both a yellow form (pH ≥ 4.4) and a red form (pH ≤ 3.2).

In the first part of this experiment, methyl orange will be prepared by the procedure outlined in the following equations. Because of the presence of acidic and basic groups, sulfanilic acid in water exists as a **zwitterion** as shown. Removal of one proton from the protonated amine with sodium carbonate followed by treatment by sodium nitrite and finally with dilute hydrochloric acid produces diazotized sulfanilic acid. The intermediate (which is only stable when kept cold) is then allowed to undergo electrophilic aromatic substitution (known as **diazonium coupling**) at the *para* position of *N,N*-dimethylaniline. The product is isolated as the yellow sodium salt.

1. What reaction would occur if an aqueous solution of diazotized sulfanilic acid were allowed to warm to room temperature?

2. At high pH the diazonium ion is largely in the form of a diazotate anion, ArN_2O^-, which is unreactive toward coupling. Coupling also fails when the pH of the solution is very low. Explain.

3. The addition of dilute hydrochloric acid to sodium nitrite produces nitrous acid, HONO, and finally NO^+. Show how the attack of NO^+ on the aromatic primary amine produces the diazonium ion.

The treatment of methyl orange (and other azo compounds) with tin(II) chloride or zinc dust in acid reduces and cleaves the azo linkage to give a mixture of the two amine fragments. For methyl orange and many other azo dyes, these are both colorless compounds. The reaction has been used as an aid in determining the structures of complex azo dyes by breaking the molecules into two more easily identified compounds.

Methyl orange $\xrightarrow[\text{H}^-]{\text{SnCl}_2 \text{ or Zn}}$ HO_3S—⬡—NH_3^+ + H_2N—⬡—$\overset{+}{N}H(CH_3)_2$

4. Write balanced equations for the reduction of methyl orange by $SnCl_2$ and Zn in acid solution.

The second type of dye molecule prepared in this experiment is malachite green, a **triphenylmethane dye**. The reaction of benzaldehyde with N,N-dimethylaniline in the presence of $ZnCl_2$ (a Lewis acid) produces a triphenylmethane derivative that can be readily oxidized to a derivative of triphenylcarbinol. Treatment of the latter with hydrochloric acid results in a stable salt, malachite green.

Leuco base of malachite green

Malachite green

5. Assume the first step in the synthesis of malachite green is

$$\text{C}_6\text{H}_5-\text{CHO} + \text{ZnCl}_2 \longrightarrow \text{C}_6\text{H}_5-\overset{+}{\text{CHOZnCl}} + \text{Cl}^-$$

Continue writing a step-by-step mechanism that will lead to the leuco base of malachite green.

Fiber Structures. Before we consider the application of dyes to fabrics, it is useful to review the chemical structures of some commonly used natural and synthetic fibers. Wool and silk are types of proteins or polypeptides in which the R groups shown in the structure can be ionic (e.g., $-\text{CH}_2\text{CO}_2^-$, $-(\text{CH}_2)_4\text{NH}_3^+$), polar (e.g., $-\text{CH}_2\text{OH}$), or nonpolar (e.g., CH_3) groups.

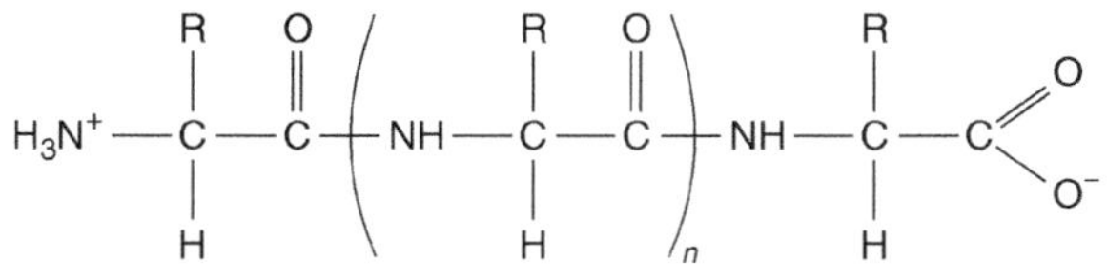

Wool or silk

Cotton is about 90% cellulose, the chief structural component of all plant cells. Conversion of the hydroxyl groups of cellulose to acetates produces cellulose triacetate, or rayon, which was one of the first synthetic fibers.

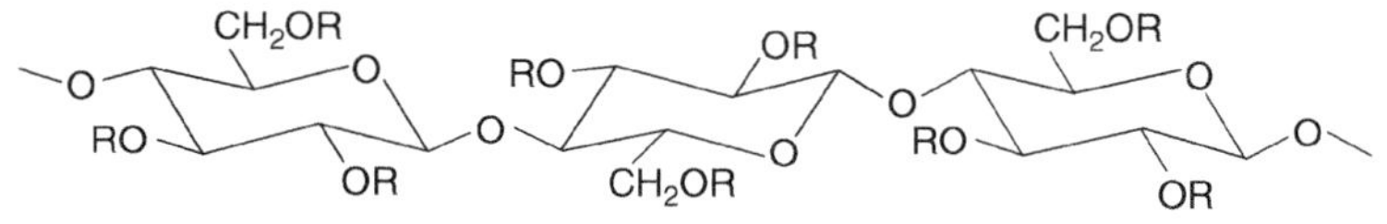

Cellulose or cotton (R = H) and cellulose triacetate or rayon (R = COCH₃)

Other common synthetic fibers include orlon (acrylic), Dacron™ (polyester), and nylon, with the structures displayed below.

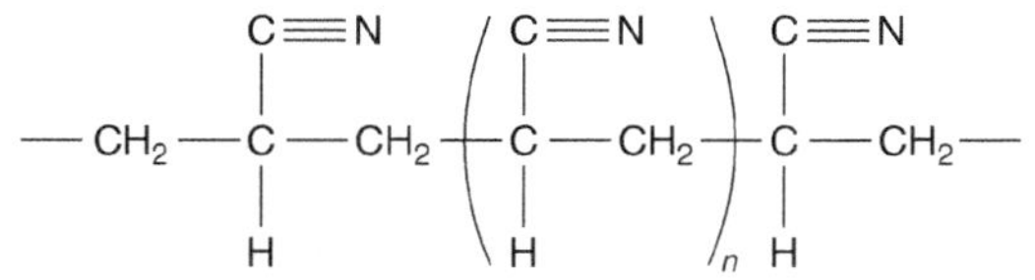

Orlon

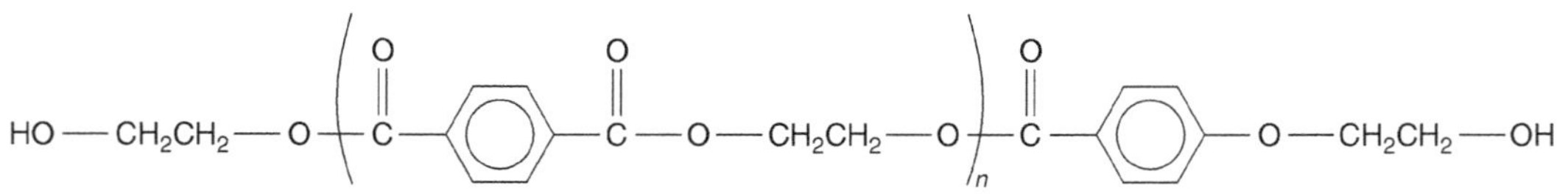

Dacron

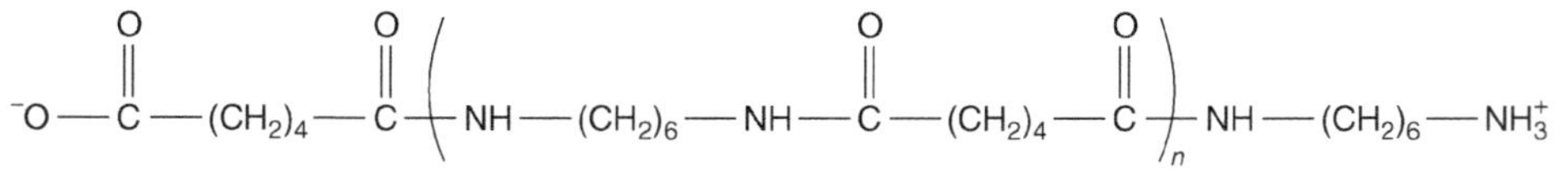

Nylon

Applying Dyes to Fabrics.   The molecular structural characteristics of various textile fibers determines the type of dye molecules or application techniques that must be used if washfast colors are to be obtained. The simplest procedure is the **direct dyeing** of fibers such as wool or silk with highly polar dyes. Both wool and silk are polypeptides that contain available carboxylic acid and amino groups. Salt formation between these groups and the acidic and basic groups on azo dyes such as methyl orange bonds the dye tightly to these fibers. **Cationic** or **basic dyes**, such as malachite green or the amino end of methyl orange, interact with $-CO_2^-$ groups in fibers, and **anionic** or **acidic dyes**, such as the sulfonic acid end of methyl orange, interact with $-NH_3^+$ groups in fibers.

$$\text{fiber} - \overset{\ominus}{CO_2} \quad \overset{\oplus}{>}N = C< \qquad \text{and} \qquad \text{fiber} - \overset{\oplus}{NH_3} \ \overset{\ominus}{O_3}S -$$

The direct dyeing of cellulose (polyglucoside) fibers such as cotton, linen, or rayon is possible, but more difficult, because polar groups on the dye molecule can hydrogen bond only weakly to the hydroxyl groups in the fiber molecule.

For many years the only method of producing washfast colors in cellulose fibers was with the aid of a **mordant** or binding agent, consisting of either a metal hydroxide or a metal complex of tannic acid applied to the cloth first. The dye is then introduced and forms an insoluble complex (sometimes called a **lake**) with the mordanted fiber. When the mordant is a heavy metal (Sn^{2+}, Fe^{3+}, etc.), it can complex to both the fiber and the dye. A phenolic dye can thus be attached to cotton by means of a metal cation mordant.

With cationic dyes such as malachite green, the mordant used is tannic acid. "Tannic acid" or "tannin" is not a single compound but a mixture of natural products found in certain barks and fruits. One common type of tannin is an ester of glucose and trihydroxybenzenecarboxylic acids, for example:

Coriligin
(a tannin)

Another method, known as **vat dying**, involves application of the dye in a water-soluble form (usually a reduced version of the final dye). A well-known example is that of indigo. In the vat process, indigo is first reduced to the leuco form with sodium hydrosulfite (sodium dithionite), NaS_2O_4.

Leucoindigo
(soluble, colorless)

Indigo
(insoluble, blue–violet)

The alkaline solutions required to dissolve the leuco form of indigo and most other vat dyes restrict their use to fabrics such a cotton or rayon, which are reasonably stable to alkaline conditions.

Other methods of dying fabrics include the use of **developed dyes**, which are formed directly in the fiber by reactions such as azo coupling; the inclusion of special functional groups in the dye molecule that will react directly with specific groups in the fiber (**fiber-reactive dyes**); and the use of water-insoluble, fiber-soluble **disperse dyes**, which will dissolve in the fiber at elevated temperatures and pressures. The latter technique is especially useful with rayon acetate, nylon, Dacron, and nonpolar polypropylene fibers.

EXERCISE

6. What is the function of strong base in increasing the water solubility of the leuco form of indigo?

A. Methyl Orange

Dissolve 250 mg (1.4 mmol) of sulfanilic acid monohydrate and 100 mg (9.4 mmol) of sodium carbonate in 5 mL of water in a 25-mL Erlenmeyer flask and cool the solution to 0°C by adding about 5 mL of cracked ice and placing the flask in an ice bath. Add a solution of 100 mg (1.45 mmol) of sodium nitrite dissolved in 1 mL of water in a small test tube or vial. Prepare a solution of 200 μL of concentrated hydrochloric acid in 1 mL of cold water in a small test tube or vial and add it slowly (in several batches) to the cold solution of sulfanilic acid with swirling and continued cooling.

CAUTION!

Many amines, such as *N, N*-dimethylaniline, are toxic and should be handled with care. Wear plastic gloves and dispense the *N, N*-dimethylaniline in a hood. Sulfanilic acid and concentrated hydrochloric acid are corrosive and should also be handled with care.

Dissolve 155 μL (149 mg, 1.23 mmol) of *N*, *N*-dimethylaniline in a solution of 250 μL of concentrated hydrochloric acid diluted with 1 mL of water in a small test tube, and cool the solution in an ice bath. Then transfer this solution slowly (in several batches) with a Pasteur pipet to the stirred, cold solution of diazotized sulfanilic acid. Some coupling will occur to produce a red color. Complete the coupling reaction and convert the resulting methyl orange to its sodium salt by *slowly* adding (Pasteur pipet; 5–6 small batches with stirring in between) 1.6 mL of 3 *M* sodium hydroxide solution.

NOTE

Coupling begins to occur as the pH approaches 7. As the color changes to yellow, add the remainder of the alkali over a period of about 1 min to allow adequate time for the coupling to occur before the pH gets too high.

Add to the reaction mixture about 2 g of clean sodium chloride,[1] add a boiling stick, and boil the suspension on a hot plate in the hood for a few minutes to digest the finely divided methyl orange and produce a microcrystalline version that is easier to filter. Cool the solution under running water and then in an ice bath. Collect the product by suction filtration on a Hirsch funnel, and wash it with a little absolute ethanol. Calculate the yield. An mp need not be determined because the compound decomposes.

1. This reduces the solubility of the methyl orange.

B. Reduction of the Azo Linkage

Dissolve about 50 mg of tin(II) chloride in 100 μL of concentrated hydrochloric acid in a small test tube, add about 10 mg of the methyl orange, and heat the mixture with a sand bath. Note the results. (As an alternative reduction procedure, treat 10 mg of methyl orange with 300–400 mg of zinc dust in 3 mL of water and 500 μL of glacial acetic acid. Warm the mixture and note the color change.)

C. Direct Dyeing with Methyl Orange

Prepare a dye solution from 500 μL of saturated sodium sulfate solution, 50 mg of methyl orange, 30 mL of water, and 1 drop of concentrated sulfuric acid (added last). Heat the beaker of solution with a hot plate or (on wire gauze on a ring) with a burner to just below the boiling point. Dip a narrow strip of multifiber fabric[2] into the solution for a few minutes. Rinse the cloth thoroughly with water and note the results.

2. Fabric containing adjacent strips of acetate rayon, cotton, nylon, Dacron,™ orlon, and wool can be obtained from Testfabrics, Inc., P.O. Box 118, 200 Blackford Ave., Middlesex, NJ 08846.

EXERCISE

7. Explain the results of your direct dyeing experiment with methyl orange in terms of the structure of the dye versus the structures of the different fibers.

NOTE

Methyl orange stains can be removed from the skin by using a paste of bleaching powder and water.

3. Use a microburner or sand bath. Use
of a hood is recommended.

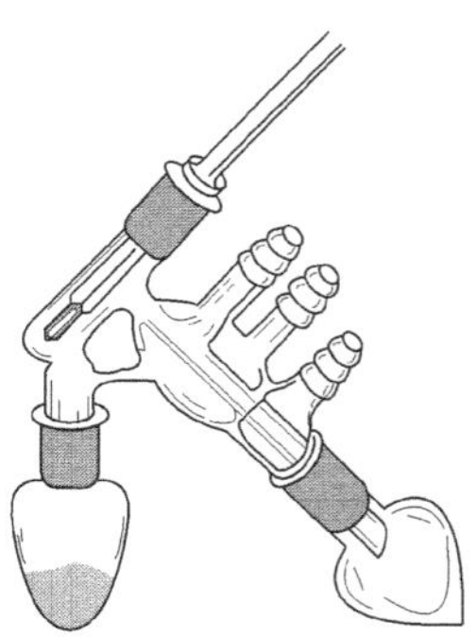

Figure E51.1 Short-path
distillation unit

D. Malachite Green

Add 400 μL each of *N,N*-dimethylaniline and benzaldehyde to a 25-mL pear-shaped
flask. Grind 200–300 mg of zinc chloride (dry granular or fused sticks) quickly (to
avoid absorption of water) and add it to the flask. Heat the flask in a sand bath to
130–140°C for 10 min.

Allow the flask to cool and then add 15 mL of water and a boiling chip. Attach
a distillation head and condenser (such as the short-path still in Figure E51.1) and distill about 10–12 mL of water (which will also remove excess *N,N*-dimethylaniline and
benzaldehyde by direct steam distillation).[3] The leuco base of malachite green remains
as a gummy blue-green mass. Allow the flask to become cool enough to touch and then
proceed.

Add 300 μL of concentrated hydrochloric acid to the leuco base and swirl the
mixture to dissolve as much as possible. Add 10 mL of water, swirl again, and add 300
mg of lead(IV) oxide. Note any changes. Shake the flask for two minutes and then add
about 300 mg of sodium sulfate to precipitate lead(II) sulfate. Heat the solution to
boiling and filter it by gravity into a beaker. Save the solution for the next part of the
experiment. To aid the cleanup, add some acetone to the residue in the flask, scrape the
inner walls to loosen lead sulfate residue, and filter this solution through the same filter
paper into a clean beaker or flask. The acetone solution can be discarded, but the filter
paper containing the lead sulfate should be disposed of as indicated in the note below.

E. Mordant Dyeing with Malachite Green

Dip a small piece of cotton cloth or a strip of the multifiber fabric into a solution of
200 mg of tannic acid in 40 mL of water and allow the cloth to remain immersed for
about 5 min. During that time dissolve 50 mg of potassium antimony tartrate (sometimes called tartar emetic) in 50 mL of water. Remove the cloth from the tannic acid
solution with a glass rod, place it in an empty beaker, and press out as much water as
possible with a spatula. Dip the cloth in the potassium antimony tartrate solution to
fasten the tannic acid to the fibers (to mordant the cloth) and again press out as much
excess water as possible from the cloth.

Heat the malachite dye solution to boiling, dip in the mordanted cloth, and continue to heat the solution for 2 min. Rinse the excess dye from the cloth with water.
Prepare a solution from 500 mg of calcium hypochlorite, 65 mL of water, and 30 mL
of saturated sodium bicarbonate solution. Dip the dyed fabric in order to bleach any

dye molecules that are not firmly attached to the fabric. (*Save this bleach solution for the indigo experiment and for cleaning stains.*) Give the fabric a final rinse with water. Try this same experiment with a piece of cotton cloth or multifiber fabric that has not been mordanted and note any differences.

NOTE

Be careful to avoid dripping the dye solution on laboratory benches or getting it on your hands or clothing. If malachite green was not prepared, the dye bath can be made by dissolving 100 mg of commercial dye in 100 mL of water. Crystal violet (or methyl violet) can be substituted.

F. Vat Dyeing with Indigo

Place 50 mg of indigo powder into a 50-mL Erlenmeyer flask, add a few drops of ethanol, and mix it into a paste with a stirring rod. Add 200 mg of sodium hydrosulfite (sodium dithionite), 2 mL of 3 M sodium hydroxide, and 20 mL of water. Stopper the flask to exclude air, shake it for several minutes, and with the stopper slightly loosened, heat the flask for a few minutes with swirling on a hot plate or sand bath. The solution will acquire a dark yellow-green color.[4] Dip a strip of cotton cloth or a strip of multifiber fabric into the solution with tweezers for 10–15 sec and hang the cloth so that the leucoindigo oxidizes in the air. After several minutes rinse the cloth with the bleach solution prepared in the malachite green experiment, wash the cloth with water, and note the results.

4. You may have to hold it up to a strong light to see this color. The surface of the solution may have an iridescent blue coating of insoluble indigo, which can be ignored.

Synthetic Organic Polymers

Although chemists are accustomed to thinking in terms of discrete molecular species with definite molecular weights, many common materials, such as elastomers, plastics, synthetic textile fibers, resins, and adhesives, do not fit this description. These materials, known as **polymers** and **macromolecules**, consist of small molecular units repeated over and over in a long chain. Molecular weights in excess of 10,000 are common for such materials, but the weights are only averages since many molecules in the polymeric material will have different chain lengths. Bonding between chains, a phenomenon known as **cross-linking**, produces rigid polymers to which the term **thermosetting** is applied because once formed they cannot be reheated in order to make them flow. **Thermoplastic** polymers are those without extensive cross-linking and can therefore be softened or melted. Many commonly encountered synthetic polymers are listed in Table E40.1.

The most extensively used commercial method for polymer production is that of **addition polymerization**. In this process monomers that contain at least one multiple bond are attached end to end to form the polymer chain. In Part **A** styrene is polymerized in this manner after first generating free radicals by thermal breakdown of dibenzoyl peroxide. Once formed, the benzoyloxy radical adds to the vinyl group of a styrene molecule to produce a benzylic radical, which can then repeat the addition process many times to form a long chain. Eventually, such a chain reaction terminates when two free radicals couple or disproportionate.

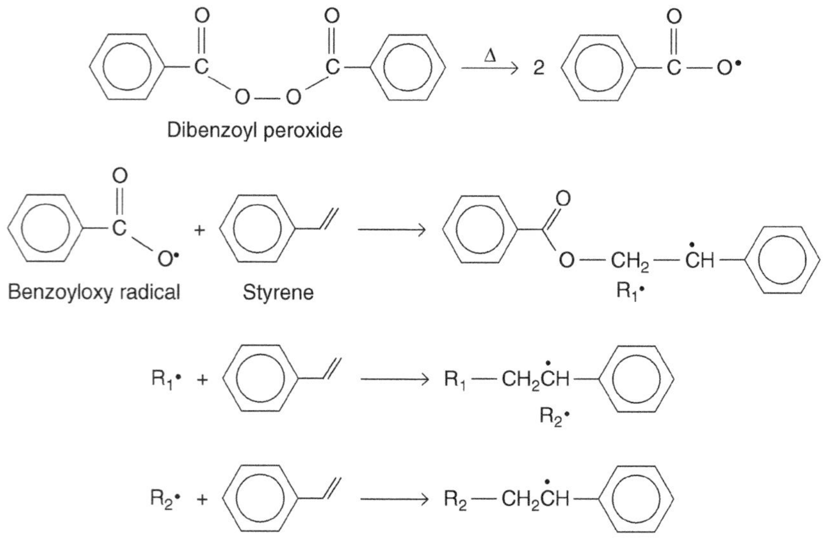

Dibenzoyl peroxide

Benzoyloxy radical   Styrene

Background Reading
Material on polymers (macromolecules) and free-radical chain reactions in any modern organic textbook

Timing
(3 h if sections are dovetailed as indicated; Part **A**: 90 min; Part **B**: 40 min; Part **C**: 20 min; Part **D**: 60 min)

144

1. Write equations for the various possible termination steps in the polymerization of styrene.

Table E40.1 Synthetic Polymers

Common or Trade Name	Recurrent Molecular Unit	Major Applications	Comments
Polyvinyl chloride (PVC)	$-CH_2-CH-$ with Cl	Molded products; piping in plumbing; guttering, tablecloths	Widely used in its softer plasticized form
Polyethylene (polythene)	$-CH_2-$	Molded products such as bottles and containers; films for packaging, building, etc.; piping	Produced in low-density (more flexible) and high-density (less flexible) forms depending on degree of branching along the chains
Polystyrene (PS)	$-CH_2-CH-$ with phenyl group	Molded products; films for electrical applications; rigid foams in building and refrigeration	
Polyvinyl acetate (PVAC)	$-CH_2-CH-$ with OAc	Emulsion paint; adhesives; butyrate ester used as the interlayer in safety glass	
Phenol-formaldehyde resins (Bakelite, phenolic resins, PF)	phenolic network structure with OH, CH_2 bridges	Molded products; paints, adhesives; laminates	This resin is almost always mixed with fillers such as wood, flour, asbestos, or mica
Polymethyl methacrylate (Perspex, Plexiglas, Lucite, Acrylic)	$-CH_2-C-$ with CH_3 and CO_2CH_3	Molded products; "unbreakable glass" sheet	
Cellulose acetate	sugar ring structure with OAc, H, and CH_2OAc groups (Triacetate shown)	Films for photographs and packaging; textile fibers	Mono-, di-, and triacetates are in common use; always used with a plasticizer

(Continued)

Table E40.1 (*Continued*)

Common or Trade Name	Recurrent Molecular Unit	Major Applications	Comments
Polyamides (Nylons)	$—(CH_2)_n CONH(CH_2)_m NHOC—$	Textile fibers; moldings	
Linear polyesters (Terylene, Dacron, Tergal, PETP)	$—R_1—\overset{\displaystyle C}{\underset{\displaystyle O}{\|\|}}—O—R_2—O—\overset{\displaystyle C}{\underset{\displaystyle O}{\|\|}}—$	Textile fibers; films	
Alkyd resins	$—R_1—\overset{C}{\|\|}—O—R_2—\overset{O}{\|}—O—\overset{C}{\|\|}—$ (with O and R_2 branch)	Paints	
Polytetrafluoroethylene (Teflon, Fluon, PTFE)	$—CF_2—$	Plastic shapes with heat and chemical resistance; plastic applications requiring low friction; coatings for cooking utensils	Very expensive
Epoxides (Araldite, EP)	(bisphenol A epoxide structure) $—\text{C}(CH_3)_2\text{—}\langle\text{aryl}\rangle—OCH_2—CH—CH_2O—$	Surface coatings; adhesives; some moldings	Expensive; various cross-linking agents used, such as amines
Silicones (SI)	$—\overset{R}{\underset{R}{Si}}—O—$	Hydrophobic protective coatings; laminates; rubbers; antifoaming agents; hydraulic fluids	Very expensive
Polyurethanes	$—R_1—NHCO_2—R_2—O_2CHN—$	Flexible foams for upholstery; rigid foams for insulation, surface coatings	
Styrene-butadiene rubber (SBR)	$—CH_2—CH{=}CH—CH_2—CH—CH_2—$ (with phenyl on CH)	Tires; general purposes	
Polychloroprene rubber (Neoprene)	$—CH_2—\overset{C}{\underset{Cl}{}}{=}CH—CH_2—$	Rubber uses requiring good weathering properties	
Butyl rubber	$—CH_2—\underset{CH_3\ \ CH_3}{C}—$	Tire inner tubes	
Polybutadiene (BR)	$—CH_2—CH{=}CH—CH_2—$	Tire treads	

2. Polymerization of styrene can also be initiated by the use of Lewis acids such as aluminum chloride. Write the equations for initiation, propagation, and termination that would be appropriate for such a process.

In addition to the polymerization of a single type of monomer unit, as in the preparation of polystyrene, two different monomer units are frequently **copolymerized** to form a chain in which both units are present in either a random arrangement or alternating. In Part **D** styrene and maleic anhydride are copolymerized in solution to form a polymer in which the maleic anhydride and styrene moieties alternate along the chain. The copolymer has rather different properties from those of a mixture of the two separate polymers formed from styrene and maleic anhydride, respectively.

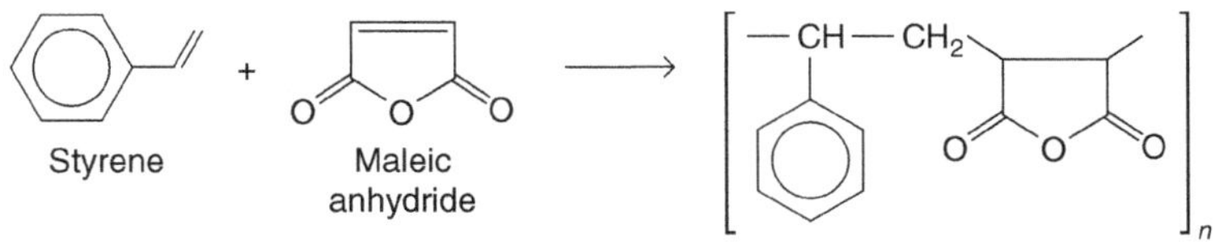

When the anhydride linkage of the styrene-maleic anhydride copolymer is hydrolyzed to the sodium salt of the corresponding diacid and applied to cellulose acetate yarns, it helps to reduce the buildup of static charge and improves fiber adhesiveness and abrasion resistance.

The experiment in Part **B** illustrates a technique known as **interfacial polymerization**, in which the polymer forms at the interface between two immiscible liquids. It also represents a second major class of polymer-forming reactions known as **condensation polymerization**, in which molecules join together with the subsequent loss of a small molecule such as water, methanol, or hydrogen chloride. In this experiment the reaction of 1,6-hexanediamine and sebacoyl chloride forms poly(hexamethylenesebacamide), a type of nylon known as nylon 6-10. The name derives from the presence of six and ten carbon atoms, respectively, in the two starting materials. The most common type of nylon used in the production of textile fibers is nylon 6-6, prepared from adipic acid and 1,6-hexanediamine. Molecular weights for this type of nylon reach values of about 15,000.

3. Dacron™ is prepared by heating a mixture of dimethyl benzene-1,
4-dicarboxylate (dimethyl terephthalate) and ethylene glycol. Write the
complete reaction and indicate the repeating unit in the polymer.

Part **C** illustrates a type of polymerization sometimes used in the preparation of
polymer foams that have excellent insulating qualities. Consider the reaction of an isocyanate with an alcohol. The product is known as a urethane and is related to both the
ester and amide functions. The treatment of an isocyanate with water produces a carbamic acid, which, unlike the thermally stable urethane, readily loses carbon dioxide
to form an amine.

$$RN{=}C{=}O + R'OH \longrightarrow RNH\overset{\displaystyle O}{\overset{\|}{C}}OR'$$

An isocyanate   A urethane

$$RN{=}C{=}O + HOH \longrightarrow RNH\overset{\displaystyle O}{\overset{\|}{C}}OH \overset{\Delta}{\longrightarrow} RNH_2 + CO_2$$

A carbamic
acid

In the preparation of polyurethane foam, the starting materials are toluene-2,4-
diisocyanate and poly(propyleneoxide)glycol. A catalyst, water, and a surface-active
agent (a detergent material) are also present. The relative quantities of the two major
components are adjusted so that there is an excess of isocyanate groups, which can
then react with a small amount of water to produce carbamic acids. It is the carbon
dioxide evolved during the decarboxylation of the carbamic acid groups that produces the foam.

Polyurethane

A. Polystyrene by Bulk Polymerization

Place about 1 mL of styrene together with enough dibenzoyl peroxide (about 10 mg)
to fit on the end of a small porcelain spatula into a 10 mm × 70 mm test tube.

Cork the tube and maintain it at about 100°C in a sand bath for at least 75 min.
During this time complete Parts **B** and **C**.

Remove the test tube from the sand bath and cool it in ice. The solid plug of polystyrene should separate from the glass. If the plug is not sufficiently loose to be tapped
out of the test tube, carefully break open the tube on the inner rim of a ceramic crock.
Being careful not to burn the material, test the melting properties of the polystyrene
by heating a small chip on the end of a metal spatula with a very small burner flame.
Grind a few small chunks of the material in a mortar and test its solubility in toluene,
acetone, and methanol.

1. Adapted from the procedure of P. W.
Morgan and S. L. Kwolek, *J. Chem.
Educ.*, **36**, 182 (1959).

B. Poly(hexamethylenesebacamide), Nylon 6-10[1]

Dissolve 1.5 mL (1.38 g, 7.0 mmol) of sebacoyl chloride in 50 mL of trichloroethylene in a 100-mL beaker. (Beakers with a diameter larger than 5 cm will not
work as well as ones of smaller diameter.) Next, dissolve 2.2 g (18.9 mmol) of
1,6-hexanediamine (or hexamethylenediamine) in 25 mL of water and carefully layer
this solution on top of the trichloroethylene. This can best be accomplished by slowly
pouring the solution along a stirring rod or through a funnel the lower end of which
is located just above the surface of the trichloroethylene. A film of nylon polymer will
form immediately at the interface between the two liquid layers.

Rubber or plastic disposable gloves are recommended for the next part of the
experiment, and it may also be desirable to place several thicknesses of newspaper on
the bench top.

With tweezers, grasp the polymer film at the center and raise it from the beaker
as a continuously forming rope that can then be collected in a second beaker. After
the reagents have been exhausted, wash the polymer several times with 50% aqueous
acetone and then with water and press it between several layers of paper towel to
remove excess moisture. Allow it to dry or dry it in an oven at about 60°C.

Take a small piece of the dry polymer on a spatula tip or in a metal spoon and
carefully heat it until it melts. With a second spatula tip or a boiling stick, see if you
can draw a fiber out of the melt. Test the solubility of the dry polymer in acetone,
toluene, and methanol.

4. Why is a molar excess of 1,6-hexanediamine required in making nylon 6-10?

5. In what way would you expect the polymer chains of a textile fiber, such as nylon 6-6, to differ from those of a rubber, such as poly(*cis*-isoprene)?

C. Polyurethane Foam

Calibrate two small paper cups so that you will be able to judge a volume of about 15 mL. Now place in one cup approximately 15 mL of component A and in the second cup about 15 mL of component B of the two polyurethane foam ingredients.[2] Pour the two viscous liquids into the bottom of a polyethylene bag that has a capacity of about 1 L and without delay thoroughly mix the components by kneading the bottom of the bag. When the components become warm and begin to foam, hold the bag at the top until the reaction is complete. Let it cool for a few minutes, then allow it to stand on the table top for about 10 min until it hardens. If desired, you can peel the polyethylene bag from around the block of urethane foam. Test a small chunk of the material for solubility in acetone, toluene, and methanol.

2. Supplied by Seemann Fiberglass, Inc., Harahan, LA, 1-800-358-1666.

D. Styrene-Maleic Anhydride Copolymer

Into a small, pear-shaped flask measure 250 μL (230 mg, 2.2 mmol) of styrene, 150 mg (1.53 mmol) of maleic anhydride, 2.5 mL of toluene, and 10 mg of dibenzoyl peroxide (handle with a porcelain spatula). Swirl the flask until the materials are dissolved, add a boiling chip, attach a condenser for reflux, and heat the reaction mixture in a sand bath at about 100°C for 15 min. A white flocculent copolymer should form in the flask.

Cool the flask in ice, add 1 mL of methanol, stir the resulting slurry, and filter it by suction on a Hirsch funnel. Wash the product with a little methanol and allow it to dry. Test the melting property of the material on a spatula tip and compare it with that of polystyrene. Test the solubility in toluene, acetone, and methanol.

6. Assuming exactly a 1:1 copolymer of maleic anhydride and styrene, calculate the yield of the copolymer based on starting maleic anhydride.

Qualitative Organic Analysis; Classification Tests

Background Reading
Relevant material in
Section 8.3 of this text

Timing
(5 h)

The classification tests that follow are a selection of the more useful tests commonly employed for the detection of aldehydes, ketones, alcohols, and phenols. It is essential that these tests be practiced on known compounds before any attempt is made to use them on unknown samples. *Although the test reagents will be available to use, methods for freshly preparing the reagents have been included.*

You will receive about 1 g (or 1 mL) of a solid or liquid unknown that is either a phenol, an alcohol, an aldehyde, or a ketone. Carry out each of the classification tests in Parts **A–G** on the suggested known samples and then on your unknown. Record the results for the unknown on the summary sheet at the end of this experiment. If two or more classification tests give conflicting results, repeat those tests carefully. Describe in your notebook the result for each test.

Purification of Liquid Samples. It is suggested that the following simple distillation be carried out on liquid unknowns prior to running the classification tests. Place about 0.5 mL of the liquid into a small test tube with a boiling chip and bring it to a boil with a microburner or sand bath. Dip the end of a Pasteur pipet into the vapor over the boiling liquid and suck a sample of vapor into the pipet. Immediate condensation will result in several drops of distilled sample.

Use of Infrared Spectroscopy. When you have finished running the classification tests on your unknown sample, obtain an IR spectrum and interpret as much of the spectrum as you can in terms of the presence or absence of possible functional groups. Compare the results of the classification tests and the infrared data.

Ketones and Aldehydes

A. 2,4-Dinitrophenylhydrazone (or 2,4-DNP) Test

The single most useful chemical test for distinguishing aldehydes and ketones from most other functional groups is the formation of a characteristic yellow to orange-red 2,4-dinitrophenylhydrazone precipitate when the sample is treated with an acidic, alcoholic solution of 2,4-dinitrophenylhydrazine.

R_2CO + O_2N—⟨benzene ring⟩—NHNH_2 ⟶ O_2N—⟨benzene ring⟩—NHN=CR_2
with NO_2 substituents

A 2,4-dinitrophenylhydrazone

The precipitate can be crystallized from 95% ethanol or from ethyl acetate and used as a derivative for purposes of identification. Among the functional groups other than aldehydes and ketones that form such a derivative are those that are transformed to an aldehyde or ketone with aqueous acid. Examples of such compounds include acetals, enamines, aldehyde hydrates, and imines; shown is an example of the equation for hydrolysis of an ethylene glycol acetal to the corresponding ketone.

$$\text{RCH(OR)}_2 \quad \text{R}_2\text{(OR)}_2 \qquad \text{RCH}=\text{CR—NR}_2 \qquad \text{RCH(OH)}_2 \qquad\qquad \text{R}_2\text{C}=\text{NH}$$

Acetals   An enamine   An aldehyde hydrate   An imine

$$\text{ethylene glycol acetal} + H_2O \xrightarrow{H_3O^+} R_2CO + HOCH_2CH_2OH$$

Derivative formation has also been noted for certain very reactive esters and anhydrides, and with some benzyl alcohols (which are oxidized by the reagent to the corresponding benzaldehydes).

EXERCISES

1. What type of spectroscopic evidence would allow a distinction to be made between ketones or aldehydes and other functional groups that might give a positive 2,4-DNP test?

2. Write the structure for the pyrrolidine enamine and for the imine of cyclohexanone. Write a balanced equation for the acid-catalyzed hydrolysis of each of these compounds.

3. On the basis of the relationship between structure and color described in Section 12.4 and the observation that both yellow and red-orange 2,4-dinitrophenylhydrazones are formed, would you expect the 2,4-DNP derivative of an aldehyde or ketone conjugated with an aromatic ring or double bond to be yellow or red-orange? Explain.

2,4-Dinitrophenylhydrazine (or 2,4-DNP) Test Reagent. Add a solution of 3 g of 2,4-dinitrophenylhydrazine in 15 mL of concentrated sulfuric acid slowly and with stirring to a mixture of 20 mL of water and 70 mL of 95% ethanol. Stir and filter.

2,4-DNP Test Procedure. A tiny drop of the pure liquid sample or a few drops from dissolving several mg of a solid sample in a minimal amount of 95% ethanol is added to several drops of test reagent on a white spot plate (or in a small test tube). If no yellow or orange-red precipitate forms immediately, allow the solution to stand at least 15 min.

Try the 2,4-DNP test on acetone, butyraldehyde, benzaldehyde, ethyl acetate, and your unknown. Describe the results and indicate what structural features will lead to a positive test. Comment on the color of any derivative.

B. Purpald[TM] Test[1]

1. Aldrich trademark for 4-amino-3-hydrazine-5-mercapto-1,2,4-triazole. The procedure is adapted from H. B. Hopps, *Aldrichemica Acta*, **33**, 28 (2000) and H. D. Durst and G. W. Gokel, *J. Chem. Educ*, **55**, 206 (1978).

The reagent reacts with both aldehydes and ketones to give the heterocyclic products shown in the first reaction below. However, only in the case of aldehydes does this initial product undergo air oxidation to give a highly colored final product. Many simple aldehydes that have some water solubility give a purple color within 1 min. However, less water-soluble aldehydes require the addition of phase transfer catalyst, Aliquat 336[TM], to help transfer the Purpald[TM] anion into the aldehyde phase. Under these latter conditions, the developing color is usually deep red to rust rather than purple. Practical grade 3-pentanone gave a positive test, but 99.5% pure ketone did not. Glacial acetic acid can give a positive test because of the presence of small amounts of acetaldehyde. Vanillin and glucose give a positive test, but require at least 10 minutes.

A companion to the Purpald[TM] test for distinguishing aldehydes and ketones is the chromic acid test, which is discussed later as one of the tests for distinguishing primary, secondary, and tertiary alcohols.

Preparing Purpald[TM] Reagent. The Purpald[TM] reagent is a solution of 2 mg/mL of reagent in $1M$ NaOH. This air-sensitive reagent must be made very fresh[2] and checked with a known aldehyde prior to each use by a student. If the reagent solution turns pale purple (aldehyde contamination), it should be replaced.

2. The solution will not last for an entire morning or afternoon lab session before giving false positive tests.

Purpald[TM] Test Procedure. Place 5–6 drops of a solution of Purpald[TM] reagent into the depression of a white spot plate. (*Avoid bubbling air through the reagent with the dropper pipet.*) Add 1 drop of a suspected liquid aldehyde or 5–10 mg of a suspected solid aldehyde and wait for 1 min. Development of a purple color indicates a positive test. If no purple color appears and the test sample appears to be insoluble, add 1 drop of Aliquat 336 (a trademark for tridecylmethylammonium chloride) so that it makes contact with the test sample. Formation of a purple or deep-red to rust color represents a positive test.

Try the test on samples of acetaldehyde, benzaldehyde, acetone, and the unknown. Describe and explain all the results in terms of the structures of the compounds.

C. Iodoform Test

In the presence of a base, most methyl ketones are converted to an enolate ion (usually in low concentration) by removal of one of the protons adjacent to the carbonyl group.

$$RCH_2CCH_3 + OH^- \overset{slow}{\rightleftharpoons} RCH_2CCH_2^- \longleftrightarrow RCH_2C{=}CH_2$$

An enolate ion

Although regioisomeric enolates form, reaction of the primary enolate with halogen is favored for steric reasons.

In the presence of a halogen (iodine is commonly used), the enolate anion and halogen react to form a monohaloketone, which—because of the electronegativity of the halogen—will form an anion more readily than the original ketone. The process is rapidly repeated until the methyl group has become a trihalomethyl group. Cleavage of the trihalo compound by base as shown then produces the salt of a carboxylic acid and a haloform. If X = iodine, water-insoluble iodoform is produced, which has a characteristic medicinal odor, is yellow in color, and melts at 119–121°C.

$$RCH_2CCH_2^- + X_2 \longrightarrow RCH_2CCH_2X + X^-$$

$$RCH_2CCH_2X + OH^- \rightleftharpoons RCH_2CCHX + H_2O \longrightarrow \longrightarrow$$

$$RCH_2CCX_3 + OH^- \rightleftharpoons RCH_2C{-}CX_3 \longrightarrow RCH_2CO_2^- + HCX_3$$

A haloform

Because of the oxidizing properties of the test solution, alkylmethylcarbinols (which are oxidized to alkyl methyl ketones) and ethanol give a positive iodoform test.

4. Would any aldehyde be expected to give a positive iodoform test? Which one(s)?

5. Can you suggest a reason why 3,5-heptanedione gives a positive iodoform test?

Preparing Iodoform Test Reagent.   Dissolve 20 g of potassium iodide and 10 g of iodine in 100 mL of water.

Iodoform Test Procedure.   Add two drops or about 50 mg of sample compound to 1 mL of water in a small vial, and if necessary, the minimum amount of 1,2-dimethoxyethane to produce a homogeneous solution. After 0.5 mL of 10% (w/v) sodium hydroxide solution has been added to the solution, the test reagent is added dropwise, with shaking, until an iodine color persists. If no yellow precipitate of iodoform appears within a few minutes, warm the mixture in a water bath at about 60°C. If the color disappears, add more of the test reagent until the color persists for at least 2 min of heating. Then add a few drops of sodium hydroxide solution to destroy the excess iodine color, dilute with water, and let stand 15 min. If yellow iodoform now precipitates, it can be recognized by its characteristic odor and the fact that it melts at 119–121°C. (A check of the mp will not be necessary unless the original sample was a pale-yellow solid.)

Try this test on cyclohexanone, acetone, isopropyl alcohol, and the unknown. Describe the results in terms of the structures of the sample compounds.

Alcohols and Phenols

D. Ferric Chloride Test

The introduction of a dilute phenol solution to a 2% solution of ferric chloride in water generally (but not always) results in the production of an intense red, orange, brown, green, blue, or violet color from the formation of a complex, the stoichiometry of which is probably as indicated.

$$3\,ArOH \;+\; [Fe(H_2O)_6]^{3+} \;\rightleftharpoons\; Fe(H_2O)_3(OAr)_3 \;+\; 3\,H_3O^+$$

Other compounds of comparable or greater acidity than phenols can also produce color changes that can interfere with the test for phenols. These include any structure, such as ethyl acetoacetate, that is capable of producing stable enols and enolate anions, and such compounds as oximes ($R_2C = NOH$) and hydroxamic acids (RCONHOH).

$$CH_3\overset{\displaystyle O}{\overset{\displaystyle \|}{C}}CH_2CO_2Et \;\rightleftharpoons\; CH_3\overset{\displaystyle OH}{\overset{\displaystyle |}{C}}=CHCO_2Et \;\rightleftharpoons\; CH_3\overset{\displaystyle O^-}{\overset{\displaystyle |}{C}}=CHCO_2Et \;+\; H^+$$

Aliphatic carboxylic acids usually produce a yellow solution, whereas aromatic carboxylic acids sometimes produce tan precipitates. However, neither of these latter results is very characteristic or useful.

A modification of the test, in which chloroform is the solvent and pyridine is added as the base, frequently gives a color change for phenols.

Preparing Ferric Chloride Reagent.   Dissolve an amount of $FeCl_3$ in water to make a 2% (by weight) solution. For the modified test, use a solution prepared by dissolving 1 g of $FeCl_3$ in 100 mL of chloroform.

Ferric Chloride Test Procedures.   Add several drops of aqueous $FeCl_3$ solution to 1 mL of a dilute (0.1–0.3% by weight) solution of the sample compound in water. Compare the color with that of several drops of reagent solution in 1 mL of water. Colors produced sometimes fade and must be watched carefully.

As a modified procedure, add 2–3 drops of the chloroform solution of ferric chloride to approximately 20 mg of solid (1 drop of liquid) dissolved or suspended in 1 mL of chloroform. Add 1 drop of pyridine.

Try the foregoing procedures on samples of phenol, ethyl acetoacetate, benzoic acid, hydroquinone (1,4-dihydroxybenzene), and the unknown. Carefully describe the results.

E. Aqueous Bromine

The exceptional electron-donating ability of the hydroxyl group allows rapid electrophilic substitution of phenols with an aqueous solution of bromine under mild conditions. The reaction usually results in bromine substitution at each available *ortho* and *para* position, with simultaneous formation of HBr, which is readily detected by the increased acidity (lower pH) of the solution. Other compounds that can give a positive test with aqueous bromine are various substituted anilines.

$$\text{(2-methylphenol)} + 2\,H_2O + 2\,Br_2 \longrightarrow \text{(2,4-dibromo-6-methylphenol)} + 2\,H_3O^+ + 2\,Br^-$$

EXERCISE

6. Phenols and alkenes will both decolorize a solution of bromine in carbon tetrachloride, yet there are simple observations that can be made during the test that will allow you to distinguish between them. Can you suggest what these observations might be?

Preparing Aqueous Bromine Reagent.   Dissolve bromine in water until a saturated solution is obtained.

Aqueous Bromine Test Procedure.   Add the aqueous solution of bromine drop by drop to a solution of approximately 20 mg (or 1 drop) of the sample in 1 mL of water. Stop when the bromine color is no longer discharged. A positive test is obtained if a sparingly soluble derivative precipitates and the solution becomes strongly acidic (as indicated by pH paper).

Try this test on phenol and the unknown. Write equations to describe the results in each case.

> **CAUTION!**
>
> Bromine or its solutions can cause serious burns on the skin and should always be handled with great care.

EXERCISE

7. Aqueous bromine converts aniline to 2,4,6-tribromoaniline and HBr. Explain why the brominated aniline does not dissolve in the acid solution as a soluble anilinium bromide salt.

F. Lucas Test (ZnCl$_2$ + HCl)

This test relies on the fact that the ease of formation of carbocations from the corresponding alcohols is highly dependent on structural features in the molecule. When other data, such as spectra, solubility properties, and negative results with other classification tests, suggest the possible presence of an alcohol, treatment with a solution of zinc chloride in hydrochloric acid can serve as a means of distinguishing between those alcohols that are quickly and easily converted to the corresponding alkyl chlorides (via the carbocation) and those that are not.

Coordination of the zinc chloride with the hydroxyl function results in the production of a sufficiently good leaving group that carbon–oxygen cleavage can now occur if a reasonably stable carbocation can form.

$$\underset{\overset{|}{H}}{R\overset{+}{O}\overset{-}{Z}nCl_2} \longrightarrow R^+ + Zn(OH)Cl_2^-$$

$$R^+ + Cl^- \longrightarrow RCl$$

If R is tertiary, benzylic, or allylic, the reaction takes place almost immediately, whereas secondary alcohols require several minutes to undergo reaction and simple primary alcohols do not react over a period of 10–15 min. The test depends on observing the formation of an insoluble liquid, alkyl chloride, as a cloudy suspension that later separates into a distinct upper layer. This requires that the alcohol be soluble in the test reagent and thus restricts the test to simple alcohols below hexyl and other alcohols that have additional functional groups that enhance the solubility in the reagent. See Table E54.1 for a summary of Lucas test results.

Use this test in conjunction with the chromic acid test described in the next section.

Preparing Lucas Reagent. Dissolve 16 g of anhydrous zinc chloride in 10 mL of concentrated hydrochloric acid with cooling.

Lucas Test Procedure. Add about 50 mg (2–3 drops) of the sample to 1 mL of the reagent in a small vial, cap, shake vigorously for a few seconds, and allow to stand at room temperature. Consult the summary table above.

If there is still doubt in distinguishing secondary from tertiary, treat the alcohol with concentrated hydrochloric acid. Only a tertiary, benzylic, or allylic alcohol will react rapidly to form the cloudy suspension of alkyl chloride.

Try this test on samples of *tert*-butyl alcohol, isoamyl alcohol (3-methyl-1-butanol), benzyl alcohol, *sec*-butyl alcohol, and the unknown.

Table E54.1 Summary of Lucas Test Results

Type of Alcohol	Time for Alkyl Halide Formation
Tertiary, allylic, benzylic	<1 min
Secondary	2–3 min
Primary	No reaction in 15 min

Table E54.2 Summary of
Chromic Acid Test Results[a]

Type of Compound	Approx. Oxidation Time
Primary alcohols	2 sec
Secondary alcohols	2 sec
Tertiary alcohols	No reaction
Aliphatic aldehydes	15 sec
Aromatic aldehydes	30–45 sec

[a] Times will vary with temperature as well as structure variation.

G. Chromic Acid Test (Jones Oxidation)

Primary and secondary alcohols are rapidly oxidized by chromium trioxide in acidic, aqueous acetone, whereas tertiary alcohols are stable to oxidation. Oxidation is readily detected by the appearance of the green Cr^{3+} ion. The test can also be used as a companion to the Purpald™ test for aldehydes, because it allows a distinction to be made between aliphatic and aromatic aldehydes that differ in the time required for oxidation. Ketones are not oxidized by these conditions. Although Table E54.2 summarizes the anticipated results, it is useful to check some authentic samples before any decisions are made based on the time required for oxidation.

EXERCISE

8. Write balanced equations for the oxidation of a primary alcohol to a carboxylic acid and a secondary alcohol to a ketone by chromic acid, H_2CrO_4.

Preparing Chromic Acid Reagent.   Add 5 g of CrO_3 in 5 mL of concentrated sulfuric acid to 15 mL of water. Allow the solution to cool.

Chromic Acid Test Procedure.   Dissolve 1 drop of liquid or 20 mg of solid sample in 1 mL of acetone in a small vial. Add 1 drop of chromic acid reagent, cap the vial, and shake it briefly.

Try this test on butyl alcohol, *sec*-butyl alcohol, *tert*-butyl alcohol, benzaldehyde, butyraldehyde, and the unknown.

Summary Form for Classification Tests

Name: ____

Unknown # ________ (aldehyde, alcohol, ketone, phenol)
2,4-DNP Test Description:

Conclusion ____
Purpald^{TM} Test Description:

Conclusion ____
Iodoform Test Description:

Conclusion ____
Ferric Chloride Test Description:

Conclusion ____
Aqueous Bromine Test Description:

Conclusion ____
Lucas Test Description:

Conclusion ____
Chromic Acid Test Description:

Conclusion ____
Infrared data on aldehyde, alcohol, ketone, or phenol unknown (spectrum attached): Indicate all important peak assignments on the spectrum.

Conclusion ____

Chemical and Spectral Analysis; Unknown Identification

Background Reading
Read Section 8.3 of the laboratory text. Review the classification tests done in the previous experiment

Timing
(10–15 h)

NOTE

Be careful not to contaminate any reagents or your unknown as you handle your sample or run various classification tests.

You will receive one solid (about 2 g) and one liquid (about 2 mL) unknown; solids will not have to be recrystallized, but liquids should be distilled. Follow the procedure outlined below and record the results on a copy of the form given at the end of this experiment and available from your stockroom.

1. Determine a capillary mp (Section 3.1) or a preliminary microscale bp (Section 3.2).
2. Distill the liquid sample with the semimicro short-path distillation apparatus pictured in Figure E55.1. Use a vacuum distillation (aspirator or diaphragm pump) if the preliminary bp was above 150°C. Determine the bp of the distilled liquid with the microscale technique outlined in Section 3.2.
3. Do the solubility tests according to the scheme in Section 8.3, but ignoring the test for solubility in concentrated sulfuric acid because simple hydrocarbons are not among the possible unknowns. If the compound is water soluble, check the solution with pH paper. You should now have the basis for distinguishing phenols and carboxylic acids from other essentially neutral compounds.
4. Do the ignition test. (Sooty flame indicates aromatic ring(s) or other carbon–carbon multiple bonds.)
5. Do the Beilstein test. (Green flame suggests halogen present, but the test is extremely sensitive.) Be certain to try known compounds with and without halogen.
6. Determine the IR spectrum of the pure unknown on Teflon™ tape. Use this spectrum and appropriate tables in Chapter 10 to confirm what functional group(s) is (are) present and absent.
7. Obtain a copy of the ^{1}H NMR and ^{13}C NMR spectra from your instructor.

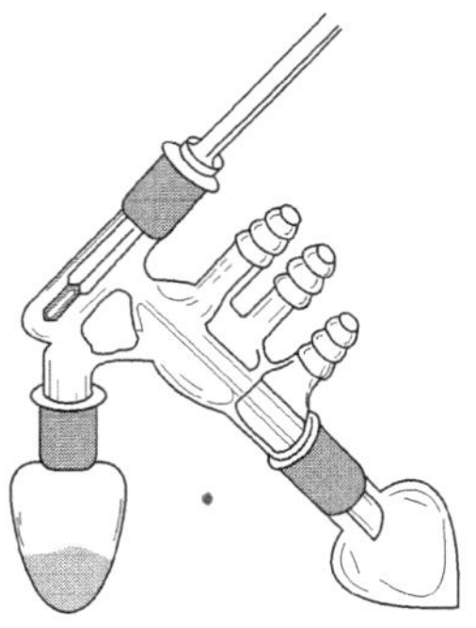

Figure E55.1   Short-path distillation unit

8. Run appropriate classification tests, but *before you try your unknown, be certain to check the reagents with known samples that will give both a positive test and a negative test.* Check the following paragraphs for which tests to run.

   - If an *aldehyde* or *ketone* is suspected, run the 2,4-DNP test, Purpald™ test, iodoform test, and chromic acid test.
   - If an *alcohol* is suspected, run the chromic acid test and Lucas test.
   - If a *phenol* is suspected, run the ferric chloride test and aqueous bromine test.

9. Use all of the previous information and an approximately 15°C range for the observed mp or bp to narrow the list of possibilities in appropriate tables available from your instructor.

10. With the structure possibilities now limited by all of the previous data, use the appropriate NMR tables in Chapter 9 to help determine or verify the structure of your unknown. Assign the peaks and multiplets to the correct structural features on the spectrum and submit your report together with your IR spectrum and the marked NMR spectrum.

<h1 align="center">Summary Form for Unknown Identification</h1>

Student Name ________________________________   Date ____________________________________

Laboratory Section ___________________________   Instructor ______________________________

Unknown #___________

Physical Constants:

mp (original sample) _________________________   mp (after optional recryst.) __________________

bp (original sample) _________________________   bp (after distillation) _______________________

lit. mp ______________________________________   lit. bp __________________________________

Ignition test: ____________________________   **Beilstein test:** ________________________

IR: Attach chart; indicate assignments for major absorptions ($4000\ cm^{-1}$ to $1300\ cm^{-1}$).

Solubility Tests: Indicate solubility to the extent of about 30 mg/mL (by + or −)

Water ____________   5% NaOH ____________   5% $NaHCO_3$ ____________   5% HCl ____________ conc. H_2SO_4
Conclusions from solubility data:

Classification Tests: (indicate pos., neg., or doubtful results by +, −, or ±, respectively.)

Test	**Result**	**Conclusion**
1.___________________	___________________	___________________
2.___________________	___________________	___________________
3.___________________	___________________	___________________
4.___________________	___________________	___________________
5.___________________	___________________	___________________
6.___________________	___________________	___________________

NMR: Finish identifying your unknown with the help of the NMR data. Draw the structure of your unknown in the space below. Assign all major peaks and multiplets in both the ^1H and ^{13}C NMR spectra by drawing an arrow with the appropriate chemical shift value to each hydrogen (or group of hydrogens) and each carbon atom in your structure.

Name of Unknown: ____
Structure of Unknown (show below):

Chemistry 204 Lab

Tests for Carbohydrates

Carbohydrates are the major source of energy in our diet, 1g of carbohydrates provides 4 Kcal of heat energy. Carbohydrate are classified into three classes, which are the monosaccharides, disaccharides and polysaccharides.

Carbohydrates have the general molecular formula CH_2O, in which hydrogen and oxygen are present in the same ratio as in water.

Types of Carbohydrates

Class	Examples/ composition	Sources
Monosaccharide	(a) Glucose------------------ (b) galactose---------------- (c) fructose -----------------	Fruit juices, honey, corn syrup Fresh peas, lactose/milk hydrolysis Fruit juices, honey
Disaccharide	(a) maltose=glucose+glucose- (b) lactose=glucose+galactose- (c)sucrose = glucose+ fructose	germinating grains,starch hydrolysis milk, yoghurt, ice-cream sugar cane, sugar beet, table sugar
Polysaccharide	(a) starch = over 50 glucoses	Rice, wheat, grains, cereals,pasta
	(b)amylose=20% starch+glucose (straight chain; joined by α- 1,4 glycosidic bonds)	Produced in plants
	(c) amylopectin=80% starch+glu (branched in 2 layers joined by α-1,4 and α-1,6 glycosic bond) .	Produced in plants
	(d) glycogen = over 50 glucoses (the most highly branched joined by α-1,4 and α-1,6 glycosidic bonds)	Stored in muscle & liver of animals
	(e) cellulose = over 50 glucoses (unbranched diagonally arranged, joined by β-glycosidic bonds)	Wood, plant leaves/roots/fiber paper,bran, beans, celery ,cotton, skins of ripe fruits like pears (only herbivores have the necessary enzymes to digest cellulose in grass)

It is occasionally necessary for biochemists to identify sugars in solutions. The chemical tests employed in these cases to distinguish among the various carbohydrates can be divided into two broad categories: (1) tests based on the production of furfural or a substituted furfural and (2) tests based on the reducing property of sugars.

TESTS BASED ON THE PRODUCTION OF FURFURAL OR A FURFURAL DERIVATIVE

When a monosaccharide is treated with a concentrated acid solution, dehydration occurs. (Disaccharides and polysaccharides are first hydrolyzed to monosaccharides by the acid medium.) If the monosaccharide is a pentose, the dehydrated product is furfural; dehydration of a hexose yields hydroxymethylfurfural.

A pentose $\xrightarrow[\text{heat}]{\text{concd. acid}}$ Furfural $+ \ 3 \ HOH$

A hexose $\xrightarrow[\text{heat}]{\text{concd. acid}}$ Hydroxymethylfurfural $+ \ 3 \ HOH$

In the presence of acid, various phenolic compounds will react with the furfural or hydroxymethylfurfural to form colored condensation products. The formation of these colored compounds constitutes a positive test for carbohydrates.

Test reagent: α-Naphthol   Orcinol   Rescorcinol
               Molisch   Bial's   Seliwanoff's

Molisch Test

The Molisch test is the most general test for carbohydrates. Compounds that are dehydrated by concentrated sulfuric acid to form furfural or hydroxymethylfurfural will react with α-naphthol (in the Molisch reagent) to yield a purple condensation product. Although this test is not specific for carbohydrates, a negative result is good evidence of the absence of carbohydrates.

167

Bial's Test

Bial's test uses orcinol to positively identify pentose sugars. It is based on the observation that furfural, which is formed from pentoses, yields a blue-green compound when treated with orcinol in the presence of ferric ions (in the Bial's reagent). The reaction is not truly specific for pentoses since other compounds, such as trioses, uronic acids, and certain heptoses, will likewise produce blue or green products. (These are typically in low concentrations in biological solutions, however.) Hexoses (yielding hydroxymethylfurfural) react with orcinol to give a yellow-brown condensation product. Therefore, observation of the color produced after addition of Bial's reagent allows one to distinguish pentoses and pentose-containing polysaccharides and nucleotides from hexoses and hexose-containing compounds.

Seliwanoff's Test

Seliwanoff's test distinguishes between ketone-containing hexoses (like fructose) and aldehyde-containing hexoses (like glucose, mannose, or galactose). Resorcinol, the reagent in Seliwanoff's test, reacts with hydroxymethylfurfural to form a red condensation product. Since ketohexoses dehydrate faster in hot hydrochloric acid to form hydroxymethylfurfural than aldohexoses, the observation of a deep red color in a relatively short period of time indicates the presence of a ketohexose. The test is particularly useful for distinguishing fructose (or a fructose-containing disaccharide) from the aldohexoses.

Procedure

Molisch Test

1 Mix 4 mL of distilled water and 2 drops of the Molisch reagent in a test tube. This tube will serve as the *control.*
2 Place 4 mL of a 1% glucose solution in a second test tube. Add 2 drops of the Molisch reagent, and mix the contents by gently shaking the test tube.
3 Incline the tube and cautiously add about 2 mL of concentrated sulfuric acid, allowing the acid to run slowly down the side of the tube. Sulfuric acid has a greater density than water and thus it will form a layer under the glucose solution. Note the color of the ring formed at the juncture of the two liquids.
4 In the same manner add sulfuric acid to the control tube. What do you observe?
5 Repeat the above test with 1% sample solutions of arabinose, fructose, lactose, sucrose, glycogen, and starch.
6 Record all your results

Bial's Test

1 Place 10 drops (0.5 mL) of each carbohydrate solution into test tubes.
2 Prepare a control test tube with 10 drops of distilled water.
3 To each tube add 15 drops (0.75 mL) of Bial's reagent (orcinol, ferric chloride, and concentrated hydrochloric acid).
4 Carefully heat each tube over a Bunsen flame until the solution just begins to boil.
5 Note the color of the product formed in each test tube. If the color is not distinct, add 25 drops (1.25 mL) of water and then 10 drops (0.5 mL) of 1-butanol and stir. The colored product will concentrate in the upper 1-butanol layer.
6 Record your results

Seliwanoff's Test

1 Place 5 drops of each carbohydrate solution into test tubes.
2 Prepare a control tube with 5 drops of distilled water.
3 To each test tube add 20 drops (1 mL) of Seliwanoff's reagent (resorcinol in dilute hydrochloric acid).
4 Place the tubes in a beaker of boiling water.
5 Record the time required for a positive test (i.e., a red color) for each sample

TESTS BASED ON THE REDUCING PROPERTY OF SUGARS

Aldoses and those disaccharides possessing a potential aldehyde group will reduce certain oxidizing reagents such as cupric ion, dinitrosalicylic acid, silver ion, and picric acid. The reaction involves the oxidation of the sugar and the corresponding reduction of the test reagent. For example, with Cu^{2+} the reaction is

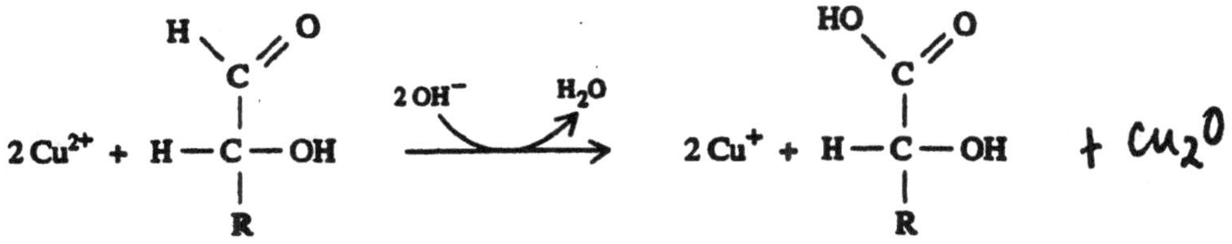

Frequently, ketoses likewise give positive results in reducing sugar tests. This is due to enolization. The presence of dilute alkali in the reaction solution promotes tautomerism of a ketose to the ene-diol which then can convert to the aldehyde.

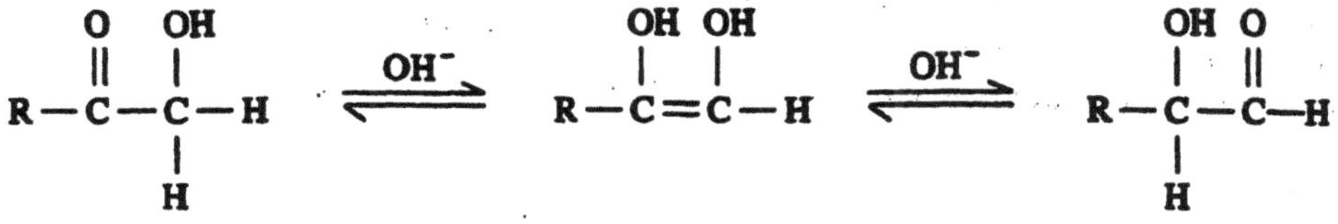

Ketose | An ene-diol | Aldose

By the same mechanism fructose is converted into glucose and gives a positive test.

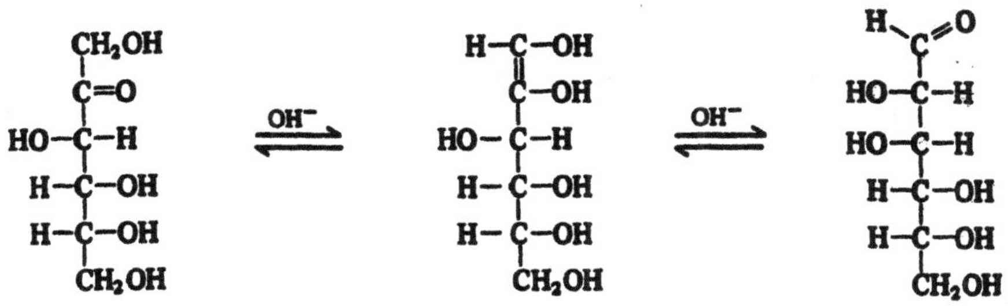

D-Fructose | Ene-diol | D-Glucose

Fehling's, Benedict's, and Barfoed's reagents all contain cupric ion in a soluble complex. In all three procedures, a positive reaction is indicated by the formation of a brick red Cu_2O precipitate. Benedict's test is generally favored over Fehling's because of the greater stability of the test reagent. It is used extensively for the detection of pathological amounts of sugar in the urine

Benedict's Test

Benedict's test is performed under mildly alkaline conditions and is a very sensitive test. The color of the precipitate may vary from green to yellow to brick red because of the varying sizes of the precipitated particles and their differing absorption of visible light. This test is by no means specific for reducing sugars. Aldehydes, in general, will reduce Benedict's reagent. Such other compounds as formic acid, hydrazobenzene, phenols, phenylhydrazine, pyrogallol, and uric acid will react in this test. The preceding compounds are not present in your sugar samples, but one or two may occur together with sugars in natural samples. For example, uric acid and sugars are both found in urine in varying amounts, depending upon the physiological condition of the individual.

Barfoed's Test

The Barfoed reagent (which contains cupric acetate in dilute acetic acid) is used to distinguish between reducing monosaccharides and disaccharides. This test differs from Benedict's test in that the oxidation–reduction reaction is carried out in an acidic rather than a basic solution. Within the same time interval (5–6 min), only monosaccharides will reduce the cupric ion in the reagent. The reaction time is determined by the rate of formation of cuprous oxide. Monosaccharides, presumably because they are smaller molecules, have the greater reactivity. If heating is prolonged, the disaccharides may be hydrolyzed by the acid and the resulting monosaccharides will give a positive test. The concentrations of the sugar solutions used in this test should be approximately the same because a more concentrated disaccharide solution will reduce the cupric ions faster than a dilute monosaccharide solution.

Procedure

Benedict's Test

1 Place 4 drops of a carbohydrate solution and 20 drops (1.0 mL) of Benedict's solution in each test tube.
2 Prepare a control tube using 4 drops of distilled water and 20 drops of the test reagent.
3 Place the tubes in a boiling water bath for 2–3 min.
4 Observe the color of the solutions and note whether a precipitate has formed. (A change in color of the solution is not indicative of a positive reaction; a precipitate must appear.)
5 Record your results

Barfoed's Test

1 Place 4 drops of each carbohydrate solution and 20 drops (1.0 mL) of Barfoed's reagent into test tubes.
2 Prepare a control tube using 4 drops of distilled water and 20 drops of the test reagent.
3 Place the tubes in a boiling water bath for 10 min.
4 Record your observations

IODINE TEST FOR POLYSACCHARIDES

The iodine test is useful for identifying starch (and occasionally other polysaccharides). It is based on the ability of the amylose helix to accommodate iodine molecules. This results in an intense blue-violet color. Branched-chain polysaccharides like amylopectin and glycogen contain less of the linear helix and, thus, react much less positively (red-brown color) with iodine. Monosaccharides and disaccharides do not react with iodine.

Iodine Test for Starch

1   Place 20 drops (1.0 mL) of each carbohydrate solution in test tubes.
2   Prepare a control tube using 20 drops of water.
3   Add 1 drop of the iodide-iodine test reagent to each test tube.
4   Record your observations

A Carbohydrate Unknown

Obtain an unknown carbohydrate from your laboratory instructor. The unknown will contain one or two of the carbohydrates listed in Table ⸱⸱ . Use the foregoing tests to determine which carbohydrate(s) you have been given. Before beginning the actual tests, you ought to develop your own scheme of analysis by means of which various possible carbohydrates may be eliminated. Hand in your results to your instructor before leaving the laboratory.

Name _______________________

Tests for Carbohydrates

Substance Tested	Molisch Test	Bial's Test	Seliwanoff's Test	Benedict's Test	Barfoed's Test	Iodine Test
Control						
Arabinose						
Glucose						
Fructose						
Lactose						
Sucrose						
Glycogen						
Starch						
Unknown						

Identity of unknown carbohydrate(s) _______________________